What This Is and How to Use It

The *Builder's Greywater Guide* describes how to work within or around building codes to install greywater systems in new construction or remodeling.

It contains new construction details and tips that will help builders successfully include greywater systems in their projects, even if they have little prior greywater experience. It also contains ammunition and strategies for presenting your case to building officials, including information on the treatment effectiveness of greywater systems.

The *Builder's Greywater Guide* is of interest to anyone getting a permit for a greywater system, installing large greywater systems or a greywater system as part of new construction or remodeling, making a greywater system for others, working on the regulation of greywater, or with an academic interest in greywater. The *Builder's Greywater Guide* is not a stand-alone work but is a companion to *The New Create an Oasis with Greywater*. If you don't have *Create an Oasis,* you'll need to obtain a copy.[1] *Create an Oasis* describes 20 types of greywater systems that work, including the design and construction of Branched Drains, our favorite system. Branched Drains are the only *simple* system both practical and legal under the California Plumbing Code (CPC) and the Uniform Plumbing Code (UPC). (In Arizona and New Mexico, you can do the even simpler free flow outlet version, arguably the most practical and simple residential reuse system of all.)

In 1991, our greywater books were aimed at environmental and legal conditions prevailing in California. Their focus has broadened considerably. *Create an Oasis* is now truly international. The *Builder's Greywater Guide* applies mostly to regulatory conditions in the US, though with a little modification, the information can be applied in almost any environmental or regulatory climate.

I see this as also as a forum for builders to benefit from each other's experiences. I sincerely hope you will contact us to let us know your experiences, so we can include them in future editions of the *Builder's Greywater Guide* (we maintain a greywater discussion group at oasisdesign.net).

Finally, the greywater field changes fast, so if you are reading this more than a year from the date of publication you may wish to check oasisdesign.net/greywater for updates.

Contents

Introduction

While there is a dramatic trend toward liberalization of greywater laws in the United States and internationally, most greywater systems in use are still illegal, and many legal systems are impractical.

Illegal use of greywater is quite common. However, it is far less obvious than, say, driving over the speed limit. Many greywater users thought they were the first to come up with the idea. In their isolation, they went on to reinvent the wheel, making the same mistakes their next-door neighbors were making in *their* isolation.

I compiled everything we could find out about practical greywater systems in the book *The New Create an Oasis with Greywater,* to which this book is a supplement. My aim is to fill an information void, enabling greywater enthusiasts to stand on the shoulders of earlier efforts and go forth from there, rather than from ground zero.

The *Builder's Greywater Guide* is the first book on greywater legality from a non-government source. It clearly illuminates greywater practices on both sides of the legality line, and on both sides of the practicality line.

Since 1989, when I started consulting on greywater system design and manufacturing greywater-safe soaps, greywater use has gone from being illegal everywhere in the United States to being legal (in some form) in most of the country. There is now a new constituency who could use a lot more information about greywater: regulators, building professionals, and owner-builders. While legality is virtually never an issue for retrofit systems, it is virtually *always* an issue for new construction or remodeling.

It is a confusing time. Many health departments are bitterly opposed to greywater reuse, yet there still has not been one documented instance of illness caused by greywater in the United States. There are similar, but significantly different, model greywater laws in the California Plumbing Code (CPC) and the Uniform Plumbing Code (UPC). They detail at least one system type which, to my knowledge, has never been built. These western model codes are full of specific mistakes, whereas the model greywater code for the East, the International Plumbing Code (IPC), makes general mistakes. Many kinks need to be worked out of these new laws. Moreover, many inspectors don't realize that greywater has been legalized at all.

In a promising new trend, Arizona and New Mexico have adopted a sensible, tiered approach to regulating greywater systems, which is totally different from the CPC, UPC, or IPC. They **don't require a permit application at all** for simple, single family systems that meet certain reasonable requirements. (Texas has followed with a similar, but slightly compromised version.)

For those who opt for systems that are both legal and practical, there are a few choices described here (unless you are lucky enough to live in Arizona or New Mexico, in which case there are many). The newly revised *Create an Oasis with Greywater* provides construction details on all systems, including the Branched Drain system, which is the one that lies most solidly in the narrow intersection of legal and practical systems at a residential scale. For others engaged in the work of making impractical designs practical, and practical designs legal, there is plenty of food for thought in these pages.

Art Ludwig

Santa Barbara, California

Is a Greywater System Appropriate?

Reasons for Builders to Install a Greywater System

Greywater systems turn "wastewater" and its nutrients into useful resources. The benefits of greywater recycling include:

❖ Reduced use of freshwater
❖ Less strain on septic tanks or treatment plants
❖ More effective purification
❖ Reduced use of energy and chemicals
❖ Groundwater recharge
❖ Increased plant growth
❖ Reclamation of nutrients
❖ Increased awareness of and sensitivity to natural cycles
❖ It's fun and the right thing to do

(*Create an Oasis* has an expanded description of these benefits in Chapter 1.) Beside these, there are benefits of particular interest to builders, described below.

Water Balance

Water supply is often a critical issue for approval or technical feasibility of a project. Reusing greywater for irrigation, especially in conjunction with conservation, rainwater harvesting, and runoff management, can get a limited water budget in the black.

Overcoming Wastewater Disposal Constraints

In those rare confluences of cooperative authorities and problem sites that are unsuited for septic tanks or sewers, greywater can be a key to getting approval for a building project which could not otherwise go forward because conventional means of water treatment would not work. This is actually specifically prohibited by most greywater laws, but it also happens occasionally in many areas (see Permit Example 1, for example).

Impact Mitigation, Green Pedigree

Greywater reuse, by lowering freshwater use for irrigation and providing spectacularly high wastewater treatment, could be an important part of a project's environmental mitigation package and Green Building pedigree.[2] Though water will almost always be saved, the overall environmental impact reduction (considering the environmental costs of the system itself) is not usually very impressive unless the reused water flow is very large or the system very simple. Resource savings are generally impressive for a facility such as a hotel with both high greywater generation and high irrigation demand. (Unfortunately, the UPC still specifically prohibits multifamily and commercial greywater installations, though the CPC now allows them. In Arizona and New Mexico, these are regulated under tier 2 or 3 of their greywater law.[3])

Reasons for Builders Not to Install a Greywater System

Owner or Users Not Supportive

Even the most user-friendly greywater system is far less idiot-proof than a sewer. Analogy: If using a sewer is like putting all your indiscriminately mixed trash in a can, using a greywater system is like hauling all your sorted valuables to the recycling center.

Greywater systems require the indulgence and intelligence of supportive players, especially the installer, at least one user, and the owner. Installing a system with only grumbling acquiescence, even if it makes total sense on the site, is asking for trouble down the road. The "I told you so" factor may magnify normal maintenance problems into perceived disasters. This caution applies doubly to those systems described as experimental.

System Too Simple

The vast majority of greywater systems in use are no more complex than a bucket dumped out the back door, or a washing machine draining into a drum with a garden hose attached to it. These systems, because they are so cheap economically and ecologically, have exceedingly attractive cost/benefit ratios and short payback times. However, they may be illegal, require constant attention, and have a short life span or other drawbacks.

Professional installation of well-designed, well-built simple systems is now allowed in Arizona and New Mexico. This is a very good thing, as the quality of the systems will surely rise. If a simple system seems best and you are in a location where professionals are not allowed to install such systems, suggest the owners do it themselves. Perhaps you can include some accommodations in the construction to make it easier for them (see Stub-Outs and Dual Waste Plumbing).

System Too Elaborate and Costly

This is an issue to keep an eye on throughout the design process—it may creep up on you. If the system has to be legal, the installation is problematic, or the owner has high standards for hands-off operation, it may turn out that the system would cost far more than the value of the water it would save.

Don't lose sight of the original point as the system grows in complexity. This concern is less acute with passive systems than with active ones. For example, it may be the case that you would need to jackhammer through a perimeter foundation to run a drain line outside by gravity. While this is a big hassle, once done, it will never require attention again. On the other hand, getting the greywater out by adding a pump pushes your client onto a slippery slope of increased maintenance and repair hassle (see Common Greywater Mistakes, *Create an Oasis*[1, 4]).

Note, however, that a life cycle analysis that takes all factors into account will generally be more favorable to greywater systems than measures such as reclaimed water plants or new water supply works. Unfortunately, many of the savings are on "external" costs which are not capturable by the people paying for the installation, such as water pollution avoided. ReWater Systems[5] has done a comprehensive life cycle analysis for their system, and also developed a spreadsheet which enables builders to do a their own cost/benefit analysis.[6]

Choosing a System

I suggest you follow the context assessment process in *Create an Oasis*, including filling out a greywater Site Assessment Form with your client. You can download a customizable version of this form from oasisdesign.net/design/consult/checklist.htm. Pay special attention to the items marked below with a ❖❖ double bullet, which should be determined early. The other information can be gathered later as needed. For a quick summary, see "Builder's Greywater Action Summary," inside back cover.

What Objectives Are You Trying to Accomplish for Your Client?

❖ ❖ Irrigate?
❖ Dispose of water safely?
❖ Other?
❖ ❖ Is it imperative that the system meet a particular economic payback timetable or is some other factor the overriding concern?

Site Features

❖ ❖ Greywater sources (quantity, quality, surge rates, and volumes) or just a list of fixtures. For a quick assessment, multiply the number of people by 300 for gallons of greywater per week, or 150 for a highly conservative household. For a more accurate assessment by fixture, see Asses Your Greywater Sources in *Create an Oasis*. (Note: Kitchen sink and dishwasher are often not legally eligible for inclusion in some areas at present.)

❖ ❖ Current and future irrigation requirements. For a very rough estimate, multiply square feet of irrigated area by ½ for gallons of water per week. For a more accurate assessment, see Appendix B, Calculating Irrigation Demand, p. 21.

❖ Existing treatment facilities and existing irrigation system

❖ Seasonal variations in water supply and water generation

❖ Access to collection plumbing

❖ ❖ Location(s) of greywater sources and location(s) of irrigation need

Site Characteristics

❖ ❖ Soil type, permeability (has there been a perk test?)

❖ ❖ Nearby surface waters

❖ Predictable disasters that may affect the design: e.g., flooding

❖ Climate (temperature and rainfall by month, wind, etc.)

❖ Groundwater level and its seasonal variation

❖ Neighborhood appropriateness

❖ ❖ Water supply constraints, e.g., economic, ecological, or availability

❖ Wastewater disposal constraints, e.g., limited septic capacity

❖ ❖ A ⅛" = 1' scale, 1' contour map of the site and a ¼" = 1' plan of the structures are very useful for greywater systems design. However, any sort of sketch is a help. The map or other description would ideally show topography, property lines, septic tanks, leach lines, wells, surface waters, buildings, major vegetation, retaining walls, drainage channels, and irrigated areas, existing and planned (much of this information may be required for a plot plan to meet CPC/UPC permit requirements; see Appendix D, Section G-4a, p. 29).

Regulatory Climate

❖ ❖ Will the project be permitted? Will it be subject to later inspection as a consequence of another project? What laws govern greywater systems in your area? Who will your inspector be and what is his/her disposition?

❖ Will other legal considerations come into play, e.g., liability exposure, environmental laws, etc.?

Client's Parameters

❖ ❖ Client's degree of enthusiasm for a greywater system

❖ ❖ Time and money constraints for maintenance, repair, and system replacement

❖ ❖ Budget (a professional installation costs a few hundred dollars for the simplest imaginable system, several hundred to several thousand for a more substantial system)

❖ Who owns the land where the project is to be built? How long are they planning to stay there? Who may follow them?

❖ Is resale value an issue?

After ❖❖-marked issues have been considered, the next question is:

What Locally Permissible Greywater System Can Best Match the Greywater Sources to the Irrigation Demand?

The answer depends a lot on the local regulator. Under the Arizona/New Mexico/Texas regulatory model (hereafter referred to as the Arizona model, since they were first), all of the greywater systems described in the System Selection Chart in *Create an Oasis* are potentially legal.

The flow chart at right yields a good answer in most situations where the CPC/UPC apply (see also the Action Summary inside the back cover).

It is all but impossible to come up with a greywater system that is simultaneously inexpensive, ecological, efficient, easy to use, and legal under the CPC/UPC. However, by sacrificing some of these parameters, you can satisfy others. There are many possible combinations of benefits and drawbacks, one or more of which will likely be a good fit for a particular situation. Your task is to determine the best fit options, then build the best one (or conclude that none are worth building in your context).

4

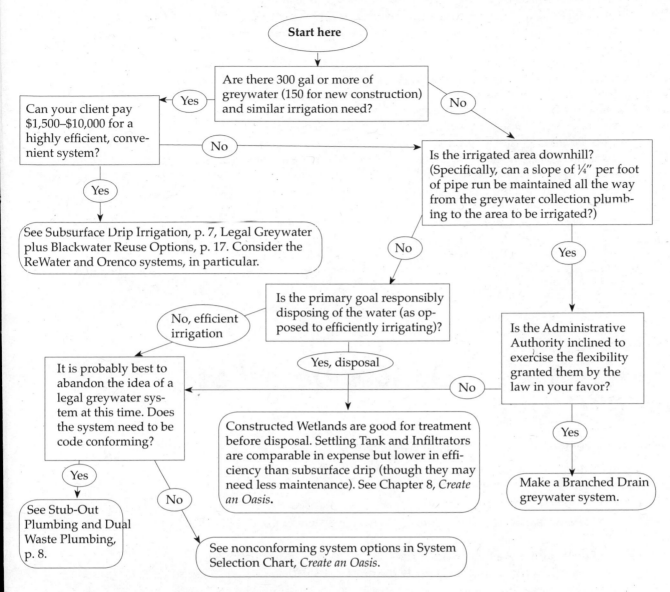

While 95%+ of homeowners shrug off the legality question for retrofit systems, for a builder this is the first factor to consider. The Arizona model is a big step in the right direction. Its biggest positive effect over the long run is likely to be from opening the door to professional installation of simple systems. The CPC/UPC in their current state contain internal contradictions, impractical requirements, and unnecessary restrictions, and require pointless extra work. They serve as a strong incentive to do an illegal system, unless the volume of greywater is sufficient to warrant Automated Sand Filtration to Subsurface Emitters, the only type of greywater system to which the law is friendly. Some loosening will likely occur in the future, so check to see if it has gotten any easier.[3]

Note on Relating with the Administrative Authority

As you attempt to reconcile your project with the requirements of the law, you may become upset. While this process can be exasperating, this paragraph is here to remind you to be unfailingly nice in your dealings with your Administrative Authority. The temptation to lash out may be compounded by the fact that they will almost certainly know next to nothing about the greywater law, as it is so seldom used. This is certainly not their fault. There is a lot of inertia to overcome, and it is largely uncharted territory. Would you like to be responsible for, say, 2½ million experimental backyard wastewater treatment facilities in Los Angeles County?

First, establish an empathic connection. What is their side of it? For example, they may feel they are putting their job on the line to help you. They will have to like you a lot to be willing to do so. Gently help them figure it out. Yours may well be the first legal system they have processed! With any luck they will reward a good attitude by cutting you some slack on the more impossible points.

Many regulators are genuinely interested in protecting public health. They might be interested to learn, for example, that the Longterm Acceptance Rate (LTAR) for greywater mulch basins is much higher than for septic leachfields (p. 14), or that using the upper reaches of the soil for purifying water protects water quality (Appendix E). If they understand these points, and understand that the code doesn't, they'll be more likely to help you on a good design that doesn't meet every letter of the code. Offer to test the system in their presence as a condition of final approval (Appendix F).

Remember, the latitude for the local Administrative Authority to go above or below state require-ments is limitless in practice. Be sure to install a system that will work, so it doesn't prejudice them against future attempts.

A note on inspections: Some greywater system installers find that inspection of the irrigation portion of their systems is cursory, while the inside plumbing is given microscopic scrutiny: "Their eyes just glaze over when they are looking at the yard, and sometimes they don't even walk out there," notes one installer. "They've never seen this stuff before. But on the collection plumbing, which they know, they rake us over the coals, hitting us because our vent extends half an inch too little above the roof, for example."

Greywater System Options Described in the CPC/UPC

Within the confines of the CPC/UPC, the options are pretty simple:

1. "Mini-Leachfields" (these bear a suspicious resemblance to full-blown septic tank leachfields and don't make sense for greywater)
2. Subsurface drip irrigation
3. Capped stub-out or dual waste plumbing for future greywater system (a great fallback position)
4. Any other means of distributing greywater subsurface that your local Administrative Authority blesses (a Branched Drain system is likely the most practical)
5. Various potentially legal blackwater reuse systems that are actually governed by septic rather than greywater code

Mini-Leachfield (Not Recommended)

Don't use this system. If you want to know why, read on.

The original Mini-Leachfield design (see *Create an Oasis*) is sound, but the legal version is a disaster. It is basically an expensive, cumbersome, disposal-only system. The irrigation efficiency could be as little as 0%. You would have to level all the lines very accurately with a transit, or have very thirsty plants to get even 20–40% irrigation efficiency. Even then, you'd have little idea of where the water was going, so it would be difficult to determine when and where supplemental freshwater irrigation was or wasn't required (you need to know this to actualize water savings). Beside, any gardener would cringe if several tons of gravel were dumped in the garden they have carefully been removing rocks from for years. Why not just add to your septic leachfield to get rid of the water? (Or do a Green Septic? Check oasisde-sign.net for the latest on this system.)

Subsurface Drip Irrigation (ReWater, Earthstar Systems)

Automated Sand Filtration to Subsurface Emitters is an excellent way to go if there are more than 300 gpd (200 for new construction) of greywater generated and corresponding irrigation demand. They are expensive and consume power, but they promise hands-off operation and have uniquely high irrigation efficiency.

Through underground drip tubing, irrigation efficiency is 80% or so, almost the same as underground drip with freshwater (during the irrigation season, that is). But any drip irrigation, especially underground, requires excellent filtration, which requires filter cleaning. Most people won't clean filters for very long. Systems with automatic backwashing filters are the best contenders in this category. You will probably not build/install this type of system yourself unless you want to go into the manufacturing business.

Jade Mountain's[7] Earthstar is a copy of a successful, smaller AGWA system which is no longer made (see The Earthstar system, next page). It may be great but hasn't been proven extensively in the field. ReWater Systems[5] manufactures an Automated Sand Filtration to Subsurface Emitters system, which distributes water through proprietary greywater distribution cones, both described in *Create an Oasis with Greywater.* This system meets the requirements for subsurface drip irrigation as per the CPC/UPC. Check our website for updates; there may be new subsurface drip contenders, particularly if these words are a few years old.

Automated Sand Filtration to Subsurface Emitters systems are appropriate for a home that generates at least 200–300 gpd of greywater. A conservative family of four generates 100 gpd or fewer, an average one maybe 200. The desirability of one of these systems at a marginal site, say, one with 150 gpd of greywater generation, depends on the cost and complexity of the installation and the value of irrigation water at that site. The system would certainly work with less than 200 gpd of greywater generation or irrigation need, but it may be a "feel good" system rather than an actual ecological benefit to the planet and economic benefit for the owner. If the quantity of greywater is small, the production of the various pumps, tanks, etc. that compose the greywater system have a greater negative environmental impact than simply wasting the greywater.

For larger flows, check out the options under Legal Greywater plus Blackwater Reuse Options, p. 17.

The Rewater System

ReWater Systems is the most solid manufacturer of this system type in the US at the moment. The ReWater computer-controlled, automatic backwashing, sand filter system handles every conceivable aspect of greywater irrigation automatically—it even coordinates freshwater irrigation. The idea is that it is your sole irrigation system (Figure 8.9, Real World Example #5, *Create an Oasis,* and Permit Example 3, p.47).

125 gpd of greywater can water a 1,000–2,000 ft² lawn through emitter cones for roughly the same cost as sprinklers: about $1.25/ft², installed. Sprinklers are cheaper on big turf areas up front, but cost more over the long term due to higher water costs from overspray and evaporation losses.

The Earthstar System

The upper capacity of the Earthstar is limited by its pipe size (2") and its requirement for manually initiated backflushing of the sand filter. This is expected every two months for an average residence. It does not have the ability to time irrigation; it sends the water out when the surge tank is full. It is $1,199, not including collection plumbing valves and parts, or irrigation components. It could be installed by a homeowner, apart from major modification to the collection plumbing. Parts only for collection plumbing and diverter valves are typically $100–$200 for a single family residence. Irrigation parts for the 800' of subsurface drip tubing typically required for a single family home with three bedrooms run about $350. Every new home on a ⅛ acre or larger lot should at least have stub-outs for one of these systems. I predict something like this will be *mandated* for new construction in California when the next major drought hits.

For supplemental freshwater irrigation, a separate above-ground drip system with a soil-moisture-sensing irrigation controller is potentially a good automated solution to actualize water savings. When the greywater (or rain) is keeping the soil moist, the controller will not initiate freshwater irrigation. If the greywater generation is inadequate, for example, because the house is unoccupied, the controller will maintain the soil moisture within bounds you specify for each zone. *Correctly installing sensors and adjusting soil-moisture-sensing irrigation controllers can be time consuming and tricky.*

The Earthstar tank and filter can be separated, but together require a 3½' x 3' area with a 15-amp ground fault interrupt (GFI) receptacle within a few feet of the slab. The greywater from all sources *must* drain to a point at least 25" above this slab by gravity. Kitchen sink water is not legally allowed or recommended by the manufacturer. The system can probably handle it, however, especially without a garbage disposal. Vegetarian kitchen sinks are less problematic due to the absence of clogging grease.

Stub-Outs and Dual Waste Plumbing

Legal stub-out plumbing is the most immediately useful achievement of the CPC (the UPC is totally silent on graywater stub-outs). Any construction you do should at least have stub-out plumbing or dual waste plumbing for a future greywater system. Pouring a slab over plumbing without greywater stub-outs is an unconscionable act. Both the hardware and laws for greywater reuse are evolving fast. It makes sense to stub out for as yet unimagined greywater systems. If only a nonconforming system makes sense, you can stub out for it legally and let the owner do the rest later.

Dual Waste Plumbing

Even if there is no legal provision for greywater in the project jurisdiction, you can accomplish fully legal provision for a future greywater system with dual waste plumbing: that is, plumb black and greywater separately until just past the point of a future greywater diversion. No law specifies where you have to join greywater and blackwater lines; outside the house is fine. (For example, if you plumbed your house as in the right hand inset of Figure 2, without the 3-way valve, that should fly.)

Stub-Outs

Greywater collection plumbing is the plumbing inside the house to either the surge tank if there is one, or a point a few feet outside the house (depending on how you define it). Stub-outs are greywater collection plumbing that dead end at a cap. They provide for easy diversion of greywater to a future greywater system to be made during the construction of a house, without having to install the complete greywater system.

In states that follow the Arizona greywater regulatory model, the collection plumbing may be the only part of the system which needs inspection. However, there isn't much guidance on how this should be done.

Stubbing out the greywater collection part of the system without the greywater distribution part of system has several advantages:

❖ Foremost, the greywater distribution system **must** be installed concurrently with the landscaping for best results. Often, the landscaping won't happen until months or years after the structure is completed and inspected.
❖ Deferring the construction of most of the system until after occupancy lowers the economic hurdle that must be cleared to attain occupancy.
❖ Greywater systems are rapidly evolving. Even if no currently available greywater system meets the owner's requirements, it makes sense to stub out greywater lines in anticipation that new system types will become available over the long life of the house. Lines entombed under a slab without stub-outs are lost to reuse forever.

Since they are a subset of builder's considerations, we'll look at greywater collection/ stub-out requirements from the inspector's viewpoint first, then we'll look at the additional considerations for builders.

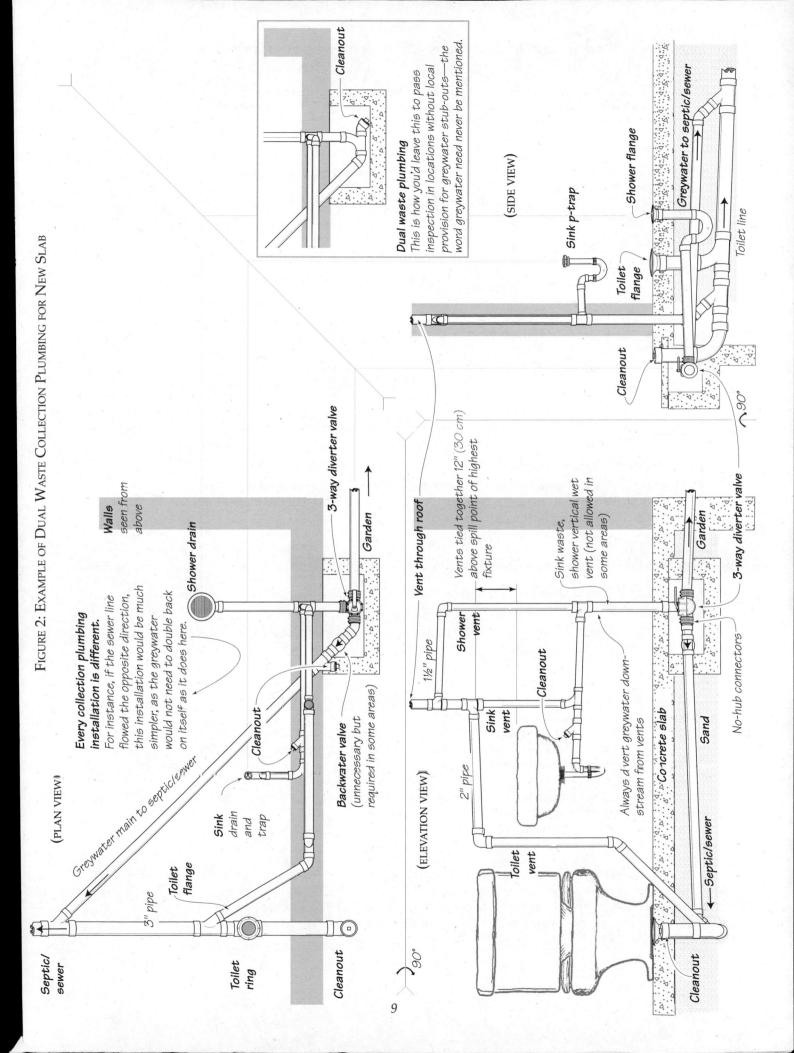

FIGURE 2: EXAMPLE OF DUAL WASTE COLLECTION PLUMBING FOR NEW SLAB

(PLAN VIEW)

Septic/sewer

Greywater main to septic/sewer

3" pipe

Toilet flange

Toilet ring

Sink drain and trap

Cleanout

Cleanout

Backwater valve (unnecessary but required in some areas)

Shower drain

3-way diverter valve

Garden →

Walls seen from above

Every collection plumbing installation is different.
For instance, if the sewer line flowed the opposite direction, this installation would be much simpler, as the greywater would not need to double back on itself as it does here.

Dual waste plumbing
This is how you'd leave this to pass inspection in locations without local provision for greywater stub-outs—the word greywater need never be mentioned.

Cleanout

(SIDE VIEW)

Sink p-trap

Toilet flange

Shower flange

Greywater to septic/sewer

Toilet line

Cleanout

⌐ 90°

(ELEVATION VIEW)

⌐ 90°

Vent through roof

Vents tied together 12" (30 cm) above spill point of highest fixture

Sink waste, shower vertical wet vent (not allowed in some areas)

1½" pipe

Shower vent

Sink vent

2" pipe

Cleanout

Always dvert greywater down-stream from vents

3-way diverter valve

Garden

Toilet vent

Concrete slab

Sand

Septic/sewer

Cleanout

No-hub connectors

9

For Regulators: How to Inspect Greywater Collection Plumbing/Stub-Outs

The Arizona Gray Water Law has only this to say about collection plumbing:

> 6. The gray water system is constructed so that if blockage, plugging, or backup of the system occurs, gray water can be directed into the sewage collection system or onsite wastewater treatment and disposal system, as applicable.

The requirements for stub-out plumbing under the CPC are minimal:

> 1. The piping has to conform to the UPC, i.e., proper slope, venting, fittings, etc., except as provided for in the greywater law.
> 2. The piping has to be permanently marked "GREYWATER STUB-OUT, DANGER—UNSAFE WATER" (p. 29, section G-5).

Not much to go on, is it? Our checklist for inspection of collection plumbing/stub-outs follows...

The ideal way to plumb the house is with greywater and blackwater totally separate, with diverter valves installed or space left for them, and the greywater lines joined to the toilet plumbing outside the house or after the point where the greywater system would later tie in. Design the layout of the future greywater system before committing to a location and height for the stub-out plumbing, if you possibly can.

Checklist for Inspection of Collection Plumbing/Stub-Outs

Required for CPC Appendix G:

❏ Stub-out is permanently marked "GRAYWATER STUB-OUT, DANGER—UNSAFE WATER" as per Appendix G, section G-5 (a)-7 (above).

Required elsewhere in plumbing codes:

❏ Pipes slope ¼" per foot minimum in all flow directions entering and leaving the diversion (the only way to do this with currently available 3-way valves is to tweak the pipes in the hubs, which do not provide for slope).
❏ Cleanouts are present every 270° of aggregate bend.

Not Mentioned in Code but Should be Required in Inspection:

❏ Diversion is downstream from vents and traps, so they will perform their function in either greywater or septic/sewer modes.

❏ In the case of a stub-out, valve is in sewer position and stub-out pipe to future greywater distribution system is securely capped.

Other Considerations

Three way valves which have a removable cover plate can function as cleanouts in all three directions.

Jandy 3-way valves, the most common type, can work equally well in all orientations; however, check that the builder has positioned the movable "inlet" designation on the valve cover to the port which is receiving the inlet water.

Not to belabor the obvious, but confirm that no toilet is connected upstream of the greywater diversion (downstream is okay, upstream connection through vent pipe connected 12" above spill point of highest fixture is allowed).

Check valve could be added with stub-out or (more commonly) later with greywater system.

Note: Check valves are not required for any other type of waste plumbing, are a source of clogging, and they form an effective trap to prevent plumbing snakes from being retracted. In practice, the check valve requirement is widely ignored by inspectors of full greywater systems for these reasons, and is rarely if ever called for during inspection of stub-outs.

One main diversion valve or multiple diversion valves to multiple outlets are both valid approaches. If your system includes kitchen sink water, this ideally should be separately divertable.

For Builders: How to Design and Construct Collection Plumbing and Greywater Stub-Outs

(There is much more on collection plumbing in *Create an Oasis*.)

One Outlet or Many?

Your first critical decision is to bring all the greywater together to one point, then divert it through one valve and distribute it from there, OR, divert greywater at multiple points with multiple valves and start with it already somewhat distributed. Once you plumb it one way or the other, you are committed. All the considerations for making this decision are covered in *Create an Oasis*.

Outlet(s) as High as Possible

The next critical design issue is to get the outlet(s) from the house as high as possible. While this can involve extra work, the value of having the outlets high can't be stressed enough. This is also covered in *Create an Oasis*.

Valve Handle(s) Accessible

Ideally the position of a diversion valve can be seen while using or on the way to use a fixture, its position can be changed while using the fixture or without going far from the fixture, and its position can be locked against meddling by children and curious guests. Sometimes this can be achieved with valve handle extensions, but often there isn't any alternative to slithering through a crawl space to change the position.

Valves Serviceable

I suggest installing the valves with no-hub connectors (which use removable clamps instead of glue) so that you can remove the entire valve for service or replacement without sawing up any pipes. If there is no space (e.g., street angles plugging right into the valve), you can use silicone sealer in place of ABS glue. This makes a watertight seal, but the fitting can easily be removed.

Valve Sources

I prefer using Jandy 3-way diverter valves, not least because the inlet can be moved to any of the three ports. Ortega and other 3-way valves also work fine. Using a tee or wye with two ball valves is commonly done, but less advisable. If the valves are seldom operated, crud can accumulate in the short dead end before the shut valve and congeal the passage shut. If you must do two ball valves, provide access for cleaning out the dead ends.

Professional Installation

I counsel greywater system do-it-yourselfers to hire a plumber to either do the collection plumbing, or check the design and your installation. There are several reasons for this:

1. Apart from the special considerations above, plumbers already know how to do collection plumbing, so you might as well take advantage of their expertise (they are generally clueless about distribution plumbing).

2. Much drain/waste/vent plumbing behavior is counterintuitive and hard to anticipate from common sense alone. Furthermore, there are real health issues with cross-connections and waste flowing in unexpected directions, more so indoors than out.

3. Collection plumbing is less apt to be changed in the future and is a longer term investment than distribution plumbing.

Seeing an impeccable installation of the collection plumbing (which they can understand) will reassure inspectors that the rest of your system (which they probably won't understand) is well thought out and executed. The converse is even more true; forget passing inspection if you've done a schlock job of the collection plumbing.

Other CPC/UPC Greywater System Options

Note: All the systems in this section are potentially, but not guaranteed, legal. They require the local Administrative Authority to exercise discretion granted them by the greywater law in your favor. Many have simpler or superior versions that are allowed under the Arizona model. These systems and variations are covered in Create an Oasis *(which also covers totally nonconforming systems).*

Section G-11 of the UPC says: "The Administrative Authority may permit subsurface drip irrigation, mini-leachfield, or *other equivalent irrigation methods* which discharge the greywater in a manner which ensures that the greywater does not surface" (emphasis added). Section G-12a appears to say that the Administrative Authority can approve anything they like. These clauses could be interpreted broadly or narrowly. All of the IPC is written broadly. A very broad interpretation would include any system which did not result in the surfacing of greywater, perhaps shown by a test of the system. A broad interpretation (on the theory that locals do not have official discretion to be more permissive, only the same or less) would be to permit any reasonable system constructed to discharge greywater at the same depth, with the same reliability, but without many of the *specific* requirements attached to either subsurface drip or Mini-Leachfields, e.g., pipe diameter, etc.

Branched Drains

For sites with continuous downhill slope from the points of greywater generation to the points of irrigation need, and flows of under 300 gpd, this design provides inexpensive, reliable, automated distribution without filter cleaning. It is critical that hard-plumbed lines have proper slope (at least ¼" per foot). All variations are legal in Arizona and New Mexico. Some have been permitted under the CPC (see Appendix D).

The first Branched Drain installations are now several years old and performing exceedingly well, even one loaded with pure unfiltered kitchen sink water. More testing is needed in heavy clay soil, freezing climates, and with greasy kitchen water, but this system looks good. There have been a handful of permitted installations, including a State of California test site in Santa Barbara. Practical Branched Drain design and construction is covered in Chapters 8, 9 and 10 of *Create an Oasis*.[1] Here is a summary of legal considerations for Branched Drain installation under the CPC/UPC.

Outlet Types

A: Free Flow Outlets (Figure 3) are the preferred option, if you've got the extra fall to spare, and you can get your inspector to see the sense in them. The advantages are that they are essentially maintenance-free and failure proof, and that inspection occurs as an incidental by-product of gardening. An inspector who is familiar with reality and human nature should be able to see how this is a great deal in trade for the effluent being visible for a lousy 2" while it is in free fall on its way into the mulch. A dogmatic inspector may have a hard time getting past this. It couldn't hurt to point out that this differs from daylighting on system failure. First, it is an intentional, vital part of the system, and the visibility itself serves an important function (for assuring monitoring of the system) and is a by-product of another functional attribute (free, unobstructable flow between pipe and basin. Second, the effluent is not ponding and festering, it is fresh, and only present for the tiny percentage of the time that water is flowing through the pipes.

B: Free Flow Outlets Concealed with Rocks are perhaps a compromise that your inspector can live with.

C: Sub-Mulch Outlets are necessary when there is not enough fall to have the outlet above the mulch or the inspector requires the outlets to be sub-surface. These will have to be inspected and occasionally mucked out to preserve even flow splitting.

D: Subsoil Infiltration Galleys are so sanitary they are fine for raw sewage or clarified septic tank effluent. They must be quite a bit larger than you think. They are overkill for most greywater, unless you have compost-laden kitchen sink water that you want to hide from rats.

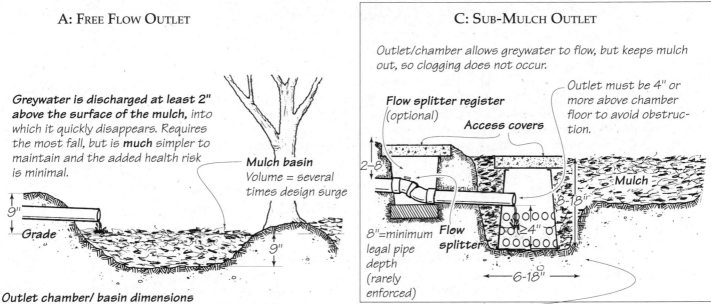

A: FREE FLOW OUTLET

Greywater is discharged at least 2" above the surface of the mulch, into which it quickly disappears. Requires the most fall, but is **much** simpler to maintain and the added health risk is minimal.

Mulch basin
Volume = several times design surge

9"
Grade
9"

C: SUB-MULCH OUTLET

Outlet/chamber allows greywater to flow, but keeps mulch out, so clogging does not occur.

Flow splitter register (optional)

Access covers

Outlet must be 4" or more above chamber floor to avoid obstruction.

2-8"

Mulch

8-18"

8"=minimum legal pipe depth (rarely enforced)

Flow splitter

≥4"

6-18"

Outlet chamber/ basin dimensions
For CPC/UPC "other" compliance, outlet can be 9" deep, same as subsurface drip.
For "Mini-Leachfield," dig down 9" below natural grade and pile up berms 9" above to get 17-18" depth required. Make them 6-18" wide. For "other," use full basin width.

B: FREE FLOW OUTLET CONCEALED WITH ROCKS

This method is intermediate between the sub-mulch and free flow options.

D: SUBSOIL INFILTRATION GALLEY

Most sanitary and most costly. All surge capacity must be met in galleys. Subsurface distribution is preferred for kitchen sink water so vermin can't use it as a food source.

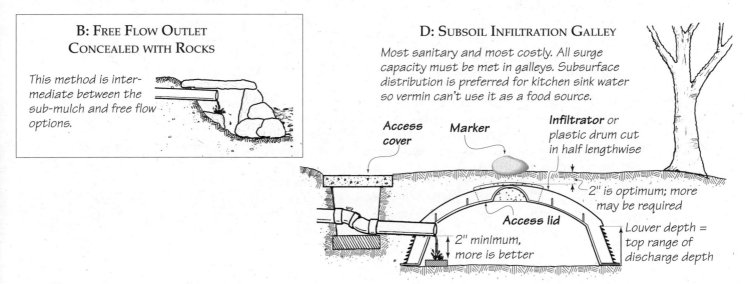

Access cover

Marker

Infiltrator or plastic drum cut in half lengthwise

2" is optimum; more may be required

Access lid

2" minimum, more is better

Louver depth = top range of discharge depth

System Sizing

Surge capacity for permitted systems must be calculated. You'll need more than you think (see Permit Example 2, CPC/UPC section G-7, and Table G-2). The short-term contribution of soil infiltration can be ignored for all but highest perk soils. The surge capacity of free flow, sub-mulch, and subsoil outlets is calculated differently.

For free flow outlets (which an enlightened inspector might pass; see Fig. 3A, above), the capacity is half the basin volume (the other half is presumed to be filled with mulch). For the depth of the basin, use the basin spill point, or the height of the preceding flow split point, whichever is lower. If the water attains the level of the preceding split, the flow will simply stop splitting and instead go on to the lower outlet.

For sub-mulch outlets, the capacity is calculated the same, but it is much more likely that the split height factor will come into play (see Fig. 3C, above). Also, to take advantage of the full basin volume, water has to be able to get out of the outlet chambers. To ease this bottleneck, use generously sized outlet chambers—5 gal pots, buckets, HDPE drums cut in half lengthwise, or infiltrators.[8]

For subsoil outlets the surge capacity is 100% of the chamber volume, as these are not filled with mulch. The depth is measured to the height of the preceding split point.

A Note on Longterm Acceptance Rate (LTAR) for Mulch Basins

Caution: The following is based on my experience with legal greywater systems in good perk soils (UPC "sandy loam"). It may carry over to systems on low perk soils, which are required to have a lower loading rate, but this has not been established.

Longterm acceptance rate (LTAR) is the longterm percolation rate of a leachfield. After three months or so, a dense biomat of bacterial slime usually forms on the soil interface. This can lower the perk rate to $\frac{1}{100}$ of its initial value. Obviously, this has tremendous implications for the required infiltration area. Codes take this into account.

As far as I know, no research has gone into distinguishing greywater mulch basin and clarified septic tank leachfield LTAR. UPC/CPC Table G-2 (greywater code) is a nearly exact copy of Table K-2, the standard for septic system leachfield area.

However, properly functioning mulch basins do not seem to form a biomat the way leachfields do. In fact, the tilling action of worms and beetles at the mulch/soil interface, and the increase in soil's organic matter over time, seem to *increase* the LTAR of a properly designed and maintained mulch basin compared to the same soil unmulched. Does this mean that code requires 100 times the necessary infiltration area for greywater to mulch basin systems? Under ideal conditions, yes. However, I believe the LTAR drops steeply when the aerobic capacity is exceeded at the soil interface of the mulch basin by high loading. This would argue for conservative design, as does the extreme variability of greywater characteristics relative to clarified septic tank effluent (consider the high BOD—Biological Oxygen Demand—of raw kitchen sink water, for example). Also, mulch basins seem to be more sensitive to initial perk rate than leachfields.

Even taking these caveats generously into account, the legally required area for greywater to mulch basin systems seems to be at least twice what is actually necessary. This is especially true for shock loads. If you size a mulch basin system at half the legal requirement, and usually have 100 gpd going into it, it can probably handle a shock load of several hundred gallons in *one hour.*

Fortunately, the CPC has dropped the requirement for multiple redundant irrigation zones, cutting the required area from four times what I think is necessary to twice. (The UPC still requires six times what's necessary.) This might be defensible for a greywater system with no alternate disposal as backup, but is unconscionable in combination with the CPC/UPC requirement for a redundant, full-sized septic/sewer system.

If it is hard to meet this unreasonable requirement, you could try the following: Make a plan which shows how you'd install the entire legally required area, *if necessary.* Explain to the Administrative Authority that LTAR for mulch basins does not drop dramatically with time, as they do not form a biomat. Offer to the inspector that, after installing half or whatever portion of the infiltration area, you'll do a test. The test could be to add a large amount of water in a short time and see if it surfaces. This could be done when the system is done (your first offer) and, if necessary, after three months of use. That is how long it takes a septic leachfield biomat to form. Carefully calculate beforehand how stringent a test it should be able to pass in theory, using the surge capacity information from System Sizing (previous section) and *Create an Oasis.* Then perform a test with half or a quarter of that volume. This amount will still generally have a generous safety factor over the expected daily loading, so the Administrative Authority can feel secure about it. Often just offering to do the performance test will relax them (see Permit Examples 1 and 2 for calculations and a sample agreement). For complete info on Branched Drain design and construction, see *Create an Oasis.*

Apparently by oversight, the CPC/UPC does not require a trap on a greywater line that does not lead to a septic tank or surge tank. Although in theory a trap is not legally required, I would not attempt to build without traps unless there was a really good reason (one architect added a trap after she found a rattlesnake in her bathtub!). But, it couldn't hurt to point out to the Administrative Authority that you are going beyond the requirements of the law.

According to the folks in Sacramento, a local jurisdiction could interpret this Branched Drain system as conforming to the CPC requirements for a Mini-Leachfield system, or the "other means of distributing greywater subsurface" clause, providing you could demonstrate that the effluent would not surface. As part of the inspection, they might require you to run a surge into the system and check for surfacing before giving final approval (see Permit Example #2). If designed to half of code or more, these systems have so much extra capacity that such a test should be passed with ease.

Equivalency with burial depth requirements and dimension ranges for the CPC/UPC Mini-Leachfields is possible but onerous (Figure 3, text below). **You're better off going for "other means" equivalency, if your inspector will allow it.** Subsurface drip only needs to be buried 9" instead of 17", for example. Fortunately, many inspectors allow wood chips in lieu of gravel (*Figure 3*, Permit Example #1). You're best off going with the "other means of distributing greywater subsurface" option. However, following are notes on the greywater laws as they might be applied *if* you have to permit your Branched Drain system under the "Mini-Leachfield" category:

Required Area (G-7)

❖ The UPC requires at least three irrigation zones (the CPC only requires one). This unreasonable requirement could possibly be filled conveniently by plumbing different fixtures to separate "family trees"; e.g., the washer waters several locations in the side yard, and the shower waters several in the back yard through a completely separate system. Otherwise, you'd have to install a diverter valve to send the water to one side (zone) of the "family tree" or the other.

❖ *Each zone to distribute all greywater produced daily without surfacing*

❖ *Meets Table G-2 design criteria of Mini-Leachfield OR*

❖ *Meets Table G-2 design criteria for subsurface drip systems*—Use Mini-Leachfield criteria to size mulch basins. These criteria will almost invariably call for a shockingly large area; see p. 14, A Note on Longterm Acceptance Rate.

Surge Tanks (G-9)

❖ Not applicable—a surge tank is not required and these systems don't need one. The CPC says "...may include surge tank." (In Colorado, one inspector required a surge tank, but allowed it to be plumbed so the water flowed straight through as if it wasn't there! In Texas, not only are surge tanks required, you're required to store water in them long enough to turn it to blackwater—probably the most unfortunate mistake in their law.)

Mini-Leachfield Systems (G-11, b)

❖ *Perforated lines minimum 3" diameter*—Perforated pipe is the outlet chamber, if you're calling it a Mini-Leachfield rather than an "other." Inspectors might be able to understand this as an infiltrator[8] or gravelless infiltration galley, an alternative to perforated pipe and gravel for septic leachfields, which they may have encountered.

❖ *High-density polyethylene pipe, perforated ABS pipe, or perforated PVC pipe*

❖ *Maximum length of perforated line-100'*—Not applicable

❖ *Maximum grade-3"/100'*—Not applicable

❖ *Minimum spacing-4'*

❖ *Earth cover of lines at least 9"*

❖ *Clean stone or gravel filter material from ¾–2½" size in trench 3" deep beneath lines and 2" above*—Use wood chips if possible

❖ *Filter fabric covers filter material*—Do not surround the leachfield completely with filter fabric or it will clog. Put filter fabric on top where gravel meets dirt, but not on the bottom. You're better off without filter fabric if you use wood chips; it just makes a mess over time.

❖ *Depth 17–18"* (G-11, b-3)—Dig down 9", pile up sides of basins 9" to get 17–18" depth required. If they think this is cheating, point out that subsurface drip only need be down 9" (G-11, a-5).

Testing (G-5, b)

❖ This is not in the law, but the Administrative Authority may require a test to show that surfacing does not occur. You may wish to suggest that your sensible design be allowed if it passes a test; this is a way of addressing both parties' concerns. One example: Submit a plan for a system which is half the required area. If a test surge shows its capacity is over, say, twice the daily design load delivered in one hour, you're done. If not, the approved plan shows the other half of the system and you go ahead and build it (see Permit Examples).

Landscape Direct

This system genre may actually legally pass under the plumbing code radar through an interpretation that because the water never enters a drain, it is not "wastewater." For example, beach showers, even in an uptight locale like Santa Barbara, just drain off over the surface into the sand. An outdoor tub might be a little more tricky. Call it a water garden, or a reflecting pool, and the showerhead a plant mister—just don't call it a greywater system.

Drum with Pump and Mesh Filter to Drip Irrigation (Not Recommended)

Do not make the common mistake of connecting a Drum with Pump and Mesh Filter to Drip Irrigation or your client will be after you for sure. To avoid callbacks, install a Drum with Effluent Pump to mulch basins or Mini-Leachfields (below), or sand filtration to subsurface emitters (p. 7).

Drums with pumps and filters were very popular during the California drought of the 1990s, but also among the most quickly abandoned systems due to the hassle of filter cleaning. Most of the handful of systems still running are those which use pantyhose filters (thrown away every couple months when they clog), and large-opening (½") emitters.

Drum with Effluent Pump and Mini-Leachfields

This proven, inexpensive system is a good all-around compromise. It was the original inspiration for the Mini-Leachfield system detailed in the CPC/UPC, which did not retain the fine attributes of the original.

This system could be constructed as described in *Create an Oasis* but with everything buried deeper, as specified in the Mini-Leachfield design criteria. This is how "equivalency" would be achieved. Greywater would come directly out ½" drip irrigation tubing (no emitters).

Drum, Pump, and Leaching Chamber or Box Trough

This system doesn't freeze easily. However, it is not especially good for reusing water. NutriCycle Systems[9] and Clivus[10] are professionally installing these in the Eastern United States. 6–18" half-pipes buried under mulch (as is done in Australia) are an inefficient but simple, robust system.

Legal Greywater plus Blackwater Reuse

All of these system options have been permitted somewhere. Ease of permitting varies widely.

There has been a recent regulatory shift in favor of treated blackwater reuse systems. These don't separate greywater, as they can handle combined wastewater. I consider this an advantage: You don't have to dual plumb, the amount of water available for reuse is 20% or so greater, and more nutrients are recycled.

Plants over Septic Tank Leachfield

Not recommended.

The UPC says you're not supposed to place woody plants near a leachfield, but of course this is done everywhere without permit problems. However, I counsel against trying this with perforated pipe. (See the cautions in *Create an Oasis.*) These are real hit or miss; a few have worked great for decades, while others are plagued with root infiltration and inability to actualize water savings. Better to do the next system.

Green Septic: Branched Drain to Infiltrators

We're working on a promising new design for reuse of septic effluent using gravelless infiltration galleys.[8] This is described in *Create an Oasis*, and you can check our website for updates.

TABLE 1: COMPARISON OF CONVENTIONAL SEPTIC AND GREEN SEPTIC SYSTEMS

Parameter	Green septic	Conventional septic
Distribution of effluent	Even, controlled	Unpredictable, concentrated
Treatment	Highly effective from aerobic and anaerobic microorganisms shallow in the soil profile; plant roots absorb nutrients	Moderately effective from anaerobic microorganisms deep in the soil profile, nutrients pollute groundwater
Water reuse	Reuse of all household greywater and blackwater facilitated by predictable, even dispersal of effluent	Impractical; some plants receive too much water, most receive none; roots can clog perforated pipes.
Inspection and service access	Provided to every part of the system	Generally little or no access
Evapotranspiration assist	Significant	Insignificant
Long term acceptance rate (LTAR)	High, due to relative absence of bacterial biomat formation, action of soil macro fauna to till soil and keep it open	Low, due to anaerobic conditions which seal the surface with a bacterial biomat (slime layer) and deter soil fauna
Action on failure	Pressure wash roots out and clean soil surface	Abandon system and make a new one elsewhere on the property

Settling Tank and Leachfields

Not recommended.

This is basically a septic tank and leachfield with plants over it, except it only has greywater going into it. The same durability cautions apply as for "plants over septic tank leachfield" (above). The only difference is that this system presumably has fewer pathogens and the lines could be a little shallower. This system is not legal under the CPC/UPC because the water is retained longer than one day. It should, however, be legal under Appendix K (septic systems). *Caution: One observer found that solids in a "greywater only" settling tank increased with time and were dispersed throughout the tank, the opposite of a septic tank receiving grey plus blackwater.*

Septic Tank, Secondary Treatment, and Subsurface Drip

These systems are widely approved under various on-site wastewater laws rather than greywater laws. There are thousands of reuse systems using subsurface drip worldwide, and some for toilet flushing. This system type is a huge economic commitment and is most common for luxury residences in environmentally sensitive areas, and larger scale systems (up to 1,000 houses).

These systems treat combined grey- and blackwater to high quality (about 1,000 coliforms/100 ml, under 10 ppm TSS—Total Suspended Solids—and low BOD). The treated effluent can be distributed through drip irrigation a few to 12" subsurface, at loading rates up to 30 gal/ft^2/day, depending on regulatory and soil conditions. (The fact that many jurisdictions allow treated *septic effluent* to be distributed only a few inches down puts the overkill 9" CPC greywater drip irrigation depth requirement into perspective.)

Costs range from $10,000–$25,000 for a turnkey system. This includes installation of a new septic tank that meets performance and durability requirements and a basic drip system with 1,500' of subsurface drip tubing.[11]

The way they keep this type of system working is with maintenance. Orenco[12], an outfit that cares about how its systems function, only sells through qualified installers and requires that you purchase a maintenance contract, forever. This is because they've found that home owners generally don't deal with septic systems. The maintenance contract costs $200–$600 a year and includes one or two annual visits to clean the pump filter, check solids, flush laterals, etc. The systems require little in the way of major maintenance: pumping out solids and pump replacement every 12–20 years. Their systems have sophisticated electronic controls. They can monitor soil moisture, dump water not needed for irrigation to alternate disposal, or add make-up freshwater if there is not enough wastewater. Orenco monitors many of their systems worldwide remotely by telemetry over radio and/or the Internet.

Constructed Wetlands

These are expensive but they're working for treatment and people are getting permits for them all over. They're not so good for irrigation reuse, as they consume much of the water themselves.

Obtain the EPA design manual.[13] Combined septic tank effluent supposedly works better than greywater alone for maintaining nutritional balance in the system. Constructed Wetlands generally require engineering, especially large systems in cold climates.[14,15]

Where To Go from Here

The rest of this book consists of appendices, including the full texts of greywater laws for most of the US, information on treatment effectiveness, etc. Beyond this, there is more information on our website, oasisdesign.net. Finally, if your situation seems to require it, we offer consulting services.

Good luck with your project!

Art Ludwig

Appendix A: Detergent Composition and Greywater Study

Office of Arid Lands Studies in Cooperation with the
Soil, Water and Plant Analysis Laboratory, University of Arizona
—Reprinted with permission ❖ Includes footnotes from Oasis—

This study was prepared for conservation-minded people who would like to use washing machine water (greywater) to irrigate their landscape plants. The list of wash-day products that follows this introduction is presented alphabetically by brand name with no endorsement of any product implied. The numbers cited should be used only as a basis of comparison among the products. It is left to the reader to choose the product(s) best suited to his/her needs. The reuse of greywater may be regulated in your area—check with your local government.

Purpose

Before greywater is used to irrigate plants, amounts of constituents potentially harmful to plants and/or soils should be known. Since labeling on detergent and other clothes-washing products often is incomplete, this study was conducted to evaluate certain product characteristics which, when introduced through greywater irrigation, may adversely affect the landscape. The specific characteristics selected for study were alkalinity, boron, conductivity, phosphate, and sodium.

Alkalinity refers to the relative amounts of alkaline chemicals in a solution. Sodium, potassium, and calcium are alkaline chemicals; they often are combined with carbonates, sulfates, or chlorides. Plants do not tolerate high concentrations of alkali salts. In soils, a buildup of alkali salts can severely reduce plant productivity.[a] In soils with high alkali concentrations, sulphur may need to be added to the soil to increase productivity.

Boron is considered a plant micronutrient, which means it is required by plants only in very, very small amounts; these usually are available in most soils. Caution: concentrations only slightly higher than those considered beneficial can cause severe injury or death to plants! The addition of boron to irrigation water should be kept at a minimum.

Conductivity is a simple measure of the amount of dissolved chemicals in a solution. These chemicals can be beneficial or harmful. The higher the conductivity, the more dissolved salts and minerals are present. In general, the higher the concentration of salts and minerals in the water, the greater the potential for adverse impacts on the environment and plant health.[b]

Phosphate is a plant food and is added to soil as a fertilizer to enhance productivity. Soils in the Tucson area typically are low in phosphate; thus, there may be some benefit to plants from the presence of this ingredient in greywater. Since phosphate has various chemical configurations, its form in detergent greywater may not be in a readily usable form to the plants and soil. This source of phosphate, therefore, should not be relied upon to assist in fertilization of plants.[c]

Sodium can act as a plant poison by changing the osmotic concentration relationship between the plant and the surrounding soil. This will reduce the plant's ability to take up water and thus will adversely impact the health of the plant. Too much sodium also destroys the structure of clay soils, making them slick and greasy by removing air spaces and thus preventing good drainage. Once a clay soil is impregnated with sodium, it is difficult to restore it to a viable condition. If soils are damaged, they may require the addition of gypsum and repeated leaching with freshwater to remove the sodium.

Although chlorine in bleach and detergents generally is expended in the washing of clothes and vaporized by the heat of hot water, some may be left in the greywater that reaches plants. If you smell chlorine during the washing process, this means that the chemical is leaving the wash water as vapor. Chlorine is considered a plant and animal poison and should not be used in the garden because it may substitute for similar nutrients, blocking normal metabolic processes. The addition of chlorine to water used for irrigation should be kept to a minimum.

Method of Analysis

All the detergents and related clothes washing products in the list below (e.g., fabric softeners) were purchased during May 1992 from various supermarkets, specialty stores, and other vendors in the Tucson, Arizona, metropolitan area.

The amount of product used in this study was based on the manufacturer's instructions for a cool-to warm-water wash in a top loading machine. The average volume of a top loading machine is 19 gallons, based on data published by Consumer Reports. Each product was dissolved in distilled/deionized water, the "cleanest" water possible, "clean" water having none or only very small amounts of dissolved salts and minerals (see table below). Tap water can contain salts and minerals in widely-varying amounts depending on its source. Using distilled/deionized water avoided addition of salts from tap water.

Discussion

Choose your detergent and clothes-washing products keeping in mind that it is better for your plants and soils to have a low alkalinity, boron, conductivity, and sodium content in the wash water. You may prefer product(s) with a higher level of one or more of these items because your clothes come out of the wash cleaner or because of personal preference.

Sandy soils are less vulnerable to damage than are clay soils because they drain better. In very low rainfall areas, apply freshwater occasionally, instead of greywater, to leach out accumuled salts.[d]

Oasis Additions:
(These comments are not part of the original paper)
[a] Potassium is a nutrient removed from soil by plants, and is thus unlikely to build up.
[b] The majority of the conductivity and alkalinity measured for plant and soil biocompatible cleaners[25] is due to potassium.
[c] According to our plant tests, phosphate in the form used in most detergents is readily usable by plants. Note, however, that phosphate is practically nonexistent in US laundry detergents now.
[d] Rainwater is comparable in quality to deionized water and is ideally suited for leaching.

Use greywater on salt-tolerant plants such as oleander, Bermuda grass, date palms, and native desert plants. Avoid using greywater on plants that prefer acid conditions, such as:[e]

Ash	Bleeding Heart (Dicentra)
Foxglove	
Philodendron	Hydrangea
Azalea	Violet
Gardenia	Camellia
Primrose	Impatiens
Begonia	Xylosma
Hibiscus	Fern
Rhododendron	Oxalis (Wood Sorrel)

The word biodegradable means that a complex chemical is broken down into simpler components through biological action. Do not be confused by the word biodegradable which often is used to imply good things. Harmful chemicals as well as beneficial ones may be biodegradable.[f]

Be aware that harmful effects are not always visible immediately and may take one to two years to appear. In any case, you should always pay attention to the health of the plants being irrigated and discontinue irrigation with greywater if signs of stress are observed.

If you choose to use greywater, we strongly recommend that you become aware of the appropriate methods to operate a greywater system and the local regulations regarding its use.

This study was prepared by the Office of Arid Lands Studies in cooperation with the Soil, Water and Plant Analysis Laboratory, University of Arizona, and is based in part on materials previously published by Pima County Cooperative Extension, University of Arizona. The study was sponsored by Tucson Water.

---

[e]From our chemical analysis, our plant studies, and user experience, it appears that the cautions below about specific plants are not a concern if you are using biocompatible cleaners.

[f]Biocompatible means that the biodegradation products are beneficial or non-harmful to a particular environment. Biocompatibility varies with the environment. For example, salt doesn't harm the ocean but is harmful for soil, phosphate is harmful for freshwater aquatic ecosystems but beneficial for soil. Most attention to date has been given to biocompatibility of cleaners with freshwater aquatic ecosystems. This study and our studies are among the first on the biocompatibility of cleaners with soil.

● Greywater-friendly detergents marked by Oasis.

● = Most biocompatible	Product Name	Product type	P/L	Conductivity at 25°C (umho/cm)	Alkalinity as CaCO3 (mg/l)	Sodium (mg/l)	Boron (mg/l)	Phosphate (mg/l)
	Ajax Ultra	Laundry	P	1130.0	219.0	292.00	0.04	11.20
●	Alfa Kleen	Det.	L	25.6	16.8	3.71	<<c	<<<d
	All	"	P	2030.0	659.0	492.00	0.10	NTe
	All Regular	"	L	116.0	29.8	39.30	<<	<<<
	Amway	"	P	939.0	310.0	227.00	<<	4.00
	Ariel Ultra	"	P	1020.0	247.0	280.00	0.03	10.80
	Arm and Hammer	"	P	2450.0	1160.0	572.00	<<	<<<
●	Bold	"	L	46.7	68.6	9.74	<<	<<<
	Bonnie Hubbard Ultra	"	P	1560.0	617.0	377.00	0.04	<<<
	Cheer Free	"	L	307.0	80.3	94.70	<<	<<<
	Cheer Ultra	"	P	710.0	149.0	171.00	0.08	<<<
	Dash	"	P	1060.0	482.0	238.00	2.14	<<<
	Dreft Ultra	"	P	737.0	328.0	189.00	9.75	<<<
	Ecovcover	"	L	132.0	63.7	24.30	<<	<<<
	ERA Plus	"	L	102.0	15.3	26.30	<<	<<<
	Fab Ultra	"	P	1140.0	199.0	443.00	<<	21.70
	Fab 1-Shot	"	Pkt	501.0	108.0	109.00	<<	5.26
	Fresh Start	"	P	510.0	106.0	132.00	0.03	8.28
	Gain Ultra	"	P	792.0	300.0	180.00	0.06	<<<
	Greenmark	"	P	1690.0	568.0	395.00	<<	1.67
	Ivory Snow	"	P	258.0	219.0	70.80	<<	NT
●	Oasis	"	L	89.6	16.2	<	<<	<<<
	Oxydol Ultra	"	P	1030.0	501.0	272.00	11.30	<<<
	Par All Temperature	"	P	2350.0	431.0	529.00	0.05	2.67
	Purex Ultra	"	P	1010.0	278.0	231.00	<<	<<<
	Sears Plus	"	P	2500.0	1200.0	635.00	<<	<<<
●	Shaklee	"	L	19.0	12.1	6.48	<<	<<<
	Shaklee Basic L	"	P	1030.0	285.0	230.00	<<	<<<
	Sun Ultra	"	P	1490.0	653.0	335.00	<<	1.58
	Surf Ultra	"	P	989.0	302.0	249.00	<<	13.70
	Tide with Bleach	"	L	329.0	58.3	95.00	2.30	<<<
	Tide Regular	"	L	291.0	61.2	93.80	0.03	<<<
	Tide Ultra	"	P	959.0	236.0	243.00	0.10	10.70
	Valu Time	"	P	1650.0	460.0	371.00	0.03	1.79
	White King	"	P	266.0	165.0	74.00	1.83	NT
	White Magic Ultra	"	P	1140.0	194.0	273.00	0.04	18.50
	Wisk Advanced Action	"	L	221.0	72.4	56.80	7.41	<<<
	Wisk Power Scoop	"	P	1160.0	360.0	319.00	<<	9.77
	Woolite	"	P	1040.0	22.3	239.00	0.17	<<<
●	Yes	"	L	42.5	10.3	6.40	<<	<<<
	Detergent Average			871.6	281.6	221.14	1.87	8.69
	Tap Water	Control	n/a	317.0	118.0	42.70	0.04	<<<
	Distilled/Deionized Water	Control	n/a	2.0	3.8	<	<<	<<<
	Snuggle Fabric Softener	Fabric Soft..	L	2.6	NT	<	<<	<<<
	Downy Fabric Softener	Fabric Soft..	L	6.4	NT	cb	<<	<<<
	Chlorox 2	Bleach	P	2880.0	1430.0	672.00	11.20	<<<
	Calgon Water Softener	Water Soft.	P	1290.0	345.0	359.00	<<	22.90

● Bio Pac

P: Powder; L: Liquid.
<: Less than the sodium detection limit of 1.0 mg/l.
<<: Less than the boron detection limit of 0.025 mg/l.
<<<: Less than the phosphate detection limit of 1.2 mg/l
NT: Testing of sample not possible.

Appendix B: Calculating Irrigation Demand

Shortcut: For average plants and average irrigation efficiency, the plant factor and irrigation efficiency cancel each other out. Most places (except deserts) have evapotranspiration rates (ET) pretty close to 1" per week. Thus, *gallons per week of irrigation demand will be approximately 0.62 times (or half) the square feet of plant area.* For a more accurate estimate, use the full formula, below:

FORMULA FOR ESTIMATING IRRIGATION DEMAND

$$D = \frac{V \times P \times A \times C}{E}$$

D = Irrigation Demand (gallons per week; *metric: liters per week*)

V = Evapotranspiration (inches per week; *metric: millimeters per week*)

P = Plant Factor (low water using = 0.3, medium = 0.5, high [e.g,. lawn] = 0.8)

E = Irrigation Efficiency (low = 0.2, average = 0.5, highest = 0.8 [subsurface drip]; see System Selection Chart in *Create an Oasis* for more system-specific values)

A = Irrigated Area (square feet; an acre is 43,560 ft². Don't count unirrigated space between plants. For a single plant, use the area under its drip line; 80 ft² for a tree with a canopy 10' in diameter, for example. *Metric: m²*)

C = Conversion factor (0.62 for inches/ft² to gallons; *1 for mm/m² to liters*)

TABLE 2: MAXIMUM EVAPOTRANSPIRATION VALUES
(INCHES PER WEEK)

Evapotranspiration increases with temperature, low humidity, and especially wind. These are *peak* values; generally you want the greywater to cover a third of the peak irrigation need. (See *Create an Oasis* for information on the relation of greywater application to ET.)

General Peak ET Values By Climate

Cool humid	0.7–1.0
Cool dry	1.0–1.4
Warm humid	1.0–1.4
Warm dry	1.4–1.8
Hot humid	1.4–2.0
Hot dry	2.0–3.2

"**Cool**" = under 70°F average midsummer high, "**Warm**"= 70°-90° average midsummer high, "**Hot**"= over 90°F. "**Humid**" = over 50% average midsummer relative humidity, "**Dry**"= under 50% average midsummer relative humidity.

Some Specific ET Values

San Francisco, CA	1.0
Santa Barbara, CA	1.3
Los Angeles, CA	1.5
San Diego, CA	1.0
Sacramento, CA	1.9

A HOMEOWNER'S GUIDE

TO

SAFE USE OF

GRAY WATER

DURING A DROUGHT

Note from Oasis:
Pressed to reveal the actual safe uses of greywater, this document is what CA DHS came up with...the unadorned truth! Their appraisal is factual and historically fascinating truth! Their appraisal is the total absence of safety margins and the total reliance on thinking users. In comprehensive. What's stunning is the total reliance on thinking users. In safety margins and the total reliance on thinking users. In stark contrast, the CPC/UPC greywater laws have safety stark contrast, the CPC/UPC greywater laws assume the worst of the factors in the double digits and assume the mid-single-digit safety user. Our designs—which, with their mid-single-digit balance between factors, strike what I feel is an optimal balance between public health, ecological, and economic considerations— look very conservative next to these recommendations.

Published by

State of California

PETE WILSON
Governor

Kenneth W. Kizer, M.D., M.P.H.
Director
Department of Health Services

March 1991

22

SAFE USE OF GRAY WATER DURING A DROUGHT

Safe use of gray water can benefit a drought-stricken community. If you are in such a community and local ordinances and policies of local agencies allow you to use gray water, this Homeowner's Guide suggests safe uses you may wish to consider.

What is Gray Water?

Gray water is used household water which has not come into contact with toilet waste, soiled diapers, or sewage. Gray water includes rinse water remaining after washing dishes by hand in a sink, used water from a bathroom sink or swimming pool, used water from a washing machine that does not receive diapers, and used water from a bathtub or shower.

Is Gray Water Safe to Use?

When care is taken, gray water can be safely used for certain important purposes around the home during a drought. It can be a valuable source of nondrinking water for the homeowner.

Why Must Care be Taken When Gray Water is Used?

A person can carry viruses, bacteria, and parasites in the intestinal tract, yet not show symptoms of the diseases they could cause. These organisms can be transferred to water in a bathtub, shower, or washing machine. Consequently, gray water should be handled with care even though it is less hazardous than sewage.

When a person who lacks immunity to intestinal disease swallows such microorganisms as a result of improper handling or use of gray water, or inhales

---

them in mist, disease may result. Thus, gray water should be handled and used in a way that avoids swallowing or inhaling of the water or residue from the gray water.

What Should Not be Included in Gray Water?

Gray water should not include water from washing of diapers, or water from a toilet bowl, or bed pan. Such water must be regarded as sewage and must be disposed of to a sewer or locally approved individual sewage disposal system (for example, a septic tank and leach field system).

Gray water should not include wash water from a kitchen sink or a dishwasher. Such water is not suitable for use because it contains grease and food particles which can clog soil, attract flies, and cause odors or other problems

What Can Gray Water be Safely Used For?

Gray water can be safely used to irrigate the following in areas that will not be contacted by children:

- Fruit trees,
- Ornamental trees and shrubs,
- Flowers and other ornamental ground cover, and
- Lawns.

If children will contact the above-cited types of areas, irrigate only by subsurface irrigation or use only the following types of gray water:

- Rinse water remaining after washing dishes by hand in a sink,

- Used water from a washing machine that does not receive diapers, or

- Used water from a bathroom sink or swimming pool.

Yikes!

You can safely irrigate a vegetable garden using used water from a washing machine that does not receive diapers, rinse water remaining after washing dishes by hand in a sink, or used water from a bathroom sink or swimming pool. Use of other gray water in a vegetable garden poses greater risk than other uses mentioned in this Homeowner's Guide. If you irrigate a vegetable garden using other gray water:

- Apply gray water only by surface irrigation, drip irrigation, or subsurface irrigation—do not spray irrigate;

- Do not irrigate a garden in a manner that allows gray water to contact or reach anything you will eat without peeling or cooking;

- Only irrigate crops that have all edible portions high above the ground surface, such as beans, corn, and tomatoes;

- Do not apply gray water near low-growing food crops, such as lettuce and strawberries; and

- Do not apply gray water near crops with subsurface portions that will be eaten raw, such as carrots, radishes, and onions.

Now!

You can safely mop floors using:

- Rinse water remaining after washing dishes by hand in a sink,

- Visibly clear rinse water from a washing machine that does not receive diapers, or

- Used water from a bathroom sink or swimming pool.

Incredible!

You can safely use gray water to water indoor ornamental plants or directly flush your toilet. Direct flushing of your toilet is done by pouring gray water from a bucket directly into the toilet bowl—not into the tank of the toilet. Do not put any gray water in the holding tank for the toilet, as a drop in water pressure might cause water from the tank to be drawn into the main water system. This could contaminate your drinking water and perhaps that of your neighbors.

Do not use gray water indoors, except as cited above.

What Other Precautions Should I Take When Using Gray Water?

Do not apply gray water in a way that forms a mist. Mist can be inhaled or swallowed.

Wash your hands immediately after touching gray water.

Keep children from playing in sprinklers that use gray water.

Do not allow gray water to pond on the surface of the ground or run off your property. Besides organisms that can cause illness, gray water may contain detergent and other chemicals which can harm aquatic life if it drains to a creek, directly or via a storm drain.

Avoid irrigation near your fence or property line that might cause gray water to seep onto your neighbor's property where it might be touched.

Gray water should not be used to wash off driveways, sidewalks, or other hard surfaces if runoff cannot be contained on your property.

Avoid any use which would allow contact between gray water and food or drinking water.

Do not put gray water into a spa or swimming pool.

Oil, grease, and high concentrations of soap should not be discharged into gray water that you plan to use.

Do not store gray water in open containers or open tanks. Open storage of gray water may cause odor problems and the attraction or breeding of insects such as mosquitoes.

Do not allow overflow or gray water from holding tanks.

If you construct or install a gray water irrigation system it will be important to adhere to a regular maintenance procedure and regularly inspect the system for leaks or blockages and verify that you are not applying more water than the soil can retain underground.

Do not overapply gray water to soil; when soil is saturated with water, gray water could pool on the surface and create a health hazard.

Remember whenever rinsing or washing diapers that this water must not be reused for any purpose and must be discarded into your sanitary sewage system.

Be thoughtful and careful in all uses of gray water. Illness will undo benefits from use of gray water.

Are There Local Requirements or Restrictions on the Use of Gray Water?

Improper alteration of your wastewater drainage connections and other features of your home plumbing system might create a cross-connection or interfere with water pressure, allowing contaminated water into your clean drinking water supply. Alteration of your home plumbing system must never be done without the approval of the local building or plumbing authority in your area.

Holding tanks, gray water treatment devices, storage systems using pumps, and other features of gray water systems may require approval of local agencies.

Consult your county health department and local building and/or plumbing authority to determine what restrictions and requirements apply in your area regarding use of gray water.

Use of toilet wastes for irrigation or fertilization is prohibited by state law that is enforced by local agencies. Disposal of human wastes by any method other than into an approved sewage system is unlawful and extremely hazardous to public health.

Is There State Regulation or State Consultation Concerning Homeowners' Use of Gray Water?

There is no state regulation or permit concerning the use of gray water. gray water treatment systems, or other gray water systems on an individual homeowner's premises. This Homeowner's Guide is the only document and consultation provided by the Department of Health Services concerning use of gray water. This guideline is not intended for use by any commercial establishment.

"We hereby wash our hands of this matter"

Can Gray Water Harm My Plants?

Gray water containing certain detergents and soaps may damage some plants and damage the ability of the soil to absorb water. Consult a commercial nursery, your county farm advisor, a licensed landscape contractor, or your local water agency.

Appendix D: Greywater Laws

Greywater is regulated locally, usually by adoption of a regional model ordinance (such as those in this Appendix), with or without modification. Some locales such as Malibu, CA have crafted their own laws from scratch.

If this sounds like it could lead to a chaotic regulatory patchwork, you're right. There are literally thousands of different greywater regulatory authorities—every state, county, and city—and no one knows how many of them are regulating greywater independently, or what exactly they're doing. However, the main regulatory systems—those that cover the majority of people in the US—are known, and follow in these pages.

This section starts with the Arizona and New Mexico greywater laws, which are the best to date.

These are followed by a checklist of CPC/UPC greywater code requirements; then the annotated full text of the CPC/UPC greywater model code for most western states; and then the International Plumbing Code (IPC), the model code for many eastern states, which has a very brief and uninformative greywater section. Each section is followed by our suggested improvements.

Arizona Greywater Law

Greywater regulation in Arizona has the following brilliant aspects:

❖ Regulators apply oversight to greywater systems in rational proportion to their possible impacts, using a three-tiered system
❖ People with low-volume, low-risk systems don't have to apply for a permit to comply with the law
❖ The law gives performance goals, not proscribed design specifics
❖ They have a short, simply worded law and a longer explanatory booklet

This is *the* model to emulate—the Arizona method makes so much sense it is hard to justify regulating greywater any other way. New Mexico has passed a similar law, Texas a somewhat similar one, and other states are considering it.

The three tiers:

1. **Systems for less than 400 gpd that meet a list of reasonable requirements** (reprinted in the next column) are all covered under a general permit without the builder having to apply for anything. With this one stroke, Arizona has raised its compliance rate from near zero to perhaps 50%. And, homeowners are more likely to work toward compliance for the informal systems that still fall short of the low bar for this first regulatory tier. What's more, the door is now open for *professionals* to install simple systems.

2. **Systems that process over 400 gpd, don't meet the list of requirements, and/or commercial, multi-family, and institutional systems** require a standard permit under the second tier.

3. **Systems over 3,000 gpd**—the third tier—are given attention by regulators on an individual basis.

The entire Arizona law for tier-one systems follows on the next page.

Annotated Arizona Greywater Law
R18-9-711. Type 1 Reclaimed Water
General Permit for Gray Water

(Strike through denotes Oasis-suggested deletions, underline denotes additions)

[From definitions:] "Graywater" means wastewater that originates from residential clothes washers, bathtubs, showers, and sinks, but does not include wastewater from ~~kitchen sinks, dishwashers and~~ toilets.

A. A Type 1 Reclaimed Water General Permit allows private residential direct reuse of gray water for a flow of less than 400 gallons per day if all the following conditions are met:

1. Human contact with gray water and soil irrigated by gray water is avoided;

2. Gray water originating from the residence is used and contained within the property boundary for household gardening, composting, lawn watering, or landscape irrigation;

3. Surface application of gray water is not used for irrigation of food plants~~, except for citrus and nut trees;~~ which have an edible portion that comes in direct contact with greywater;

4. The gray water does not contain hazardous chemicals derived from activities such as cleaning car parts, washing greasy or oily rags, or disposing of waste solutions from home photo labs or similar hobbyist or home occupational activities;

5. The application of gray water is managed to minimize standing water on the surface, for example, by splitting the flow, moderate application rates, and generous mulching;

6. The gray water system is constructed so that if blockage, plugging, or backup of the system occurs, gray water can be directed into the sewage collection system or onsite wastewater treatment and disposal system, as applicable (except as provided for under 10, below). The gray water system may include a means of filtration to reduce plugging and extend system lifetime;

7. Any gray water storage tank is covered to restrict access and to eliminate habitat for mosquitoes or other vectors;

8. The gray water system is sited outside of a floodway;

9. The gray water system is operated to maintain a minimum vertical separation distance of at least five feet from the point of gray water application to the top of the seasonally high groundwater table;

10. For residences using an onsite wastewater treatment facility for black water treatment and disposal, the use of a gray water system does not change the design, capacity, or reserve area requirements for the onsite wastewater treatment facility at the residence, and ensures that the facility can handle the combined black water and gray water flow if the gray water system fails or is not fully used. Alternatively, the greywater system shall be designed with two valved zones, each of which can accommodate the full expected greywater volume. Providing the greywater system passes a flow test in each zone, the capacity of the on-site system may be reduced, or in the instance that an approved composting toilet system is present, eliminated;

11. Any pressure piping used in a gray water system that may be susceptible to cross connection with a potable water system clearly indicates that the piping does not carry potable water;

12. Gray water applied by surface irrigation does not contain water used to wash diapers or similarly soiled or infectious garments unless the gray water is disinfected before irrigation; and

13. Surface irrigation by gray water is only by flood or drip irrigation. Containment within horticultural basins or swales is encouraged for flood irrigation;

14. It is required that kitchen sink water be applied subsoil or contained within a rat-proof outlet shield;

15. Greywater diverter valves should be downstream from traps and vents in plumbing that leads to septic or sewer.

B. Prohibitions. The following are prohibited:

1. Gray water use for purposes other than irrigation, and

2. Spray irrigation.

C. Towns, cities, or counties may further limit the use of gray water described in this Section by rule or ordinance.

The main feedback to the Arizona Department of Environmental Quality (DEQ) has been from environmentalists upset that greywater plus composting toilets are not allowed.

The DEQ may revise the rules in the future to allow kitchen sink water. This would solve the composting toilet issue if item 10 was also revised so it didn't call for a full-sized septic.

There is much more on the Arizona law in our Greywater Policy Center, oasisdesign.net/greywater/law.

New Mexico Greywater Law

The New Mexico greywater law is similar to the Arizona version, though not quite as good. This is the meat of it, with our suggested improvements in _underline_ and ~~strike-thru~~:

Section 1. Section 74-6-2 NMSA 1978 (being Laws 1967, Chapter 190, Section 2, as amended) is Amended to Read:

...L. shall not require a permit for applying less than ~~two hundred fifty~~ _four hundred_ gallons per day of private residential gray water originating from a residence for the resident's household gardening, composting or landscape irrigation if:

1. a constructed gray water distribution system provides for overflow _and/or diversion_ into the sewage collection or on-site wastewater treatment and disposal system;

2. a gray water storage tank is covered to restrict access and to eliminate habitat for mosquitoes or other vectors;

3. a gray water system is sited outside of a floodway;

4. gray water is vertically separated at least five feet above the groundwater table;

5. gray water pressure piping is clearly identified as a nonpotable water conduit;

6. gray water is used on the site where it is generated and does not run off the property lines;

7. ponding is prohibited, application of gray water is managed to minimize standing water on the surface and standing water does not remain for more than twenty-four hours;

8. gray water is not sprayed; and

9. gray water use within municipalities or counties complies with all applicable municipal or county ordinances enacted pursuant to Chapter 3, Article 53 NMSA 1978

This law would benefit from the same improvements suggested for the Arizona law, previous page.

CPC/UPC Legal Requirements

Summary

See the annotated text of the California greywater law (following) for details (G-section references). Note that the CPC's Appendix G (applied in California and included here) differs from the UPC's Appendix G, which is what your inspector will find in his/her UPC code book. The letter designation also may change when the code is revised. This summary is based on 2000 codes. Our suggested changes follow the code.

(GW = greywater, GWS = greywater system)

❖ GW used only for subsurface landscape irrigation (G-1a)
❖ GWS now allowed for commercial/multifamily in CA (recent change) (G-1a)
❖ No connection to potable water system (G-1a)
❖ No GW surfacing (G-1a)
❖ UPC applies to GWS except as provided in Appendix G CPC (G-1a)
❖ No part of GWS may be on a lot other than the one which generated the GW (G-2c)
❖ Location of components must comply with minimum distances in Table G-1 (G-1c, G-1f, Table G-1)
❖ Plot plan to scale with all information in Section G-4 (a) required for submittal (summarized under first item of GW Measures Checklist, below, and in G-1d, G-4a)
❖ GW can't discharge where it could increase the likelihood of a landslide (G-4e)
❖ Other disposal system (septic, sewer not mentioned) can't be compromised or reduced in size on account of the GWS (G-1f)
❖ Installers must provide users with an operation and maintenance manual (G-1g). Manuals should be supplied with commercial systems. (For a non-manufactured system, perhaps a copy of _Create an Oasis_, or the California Department of Water Resources _Graywater Guide_,[16] plus some comments on the particular installation would suffice.)
❖ GWS cannot accept GW from kitchen sink, dishwashers (check for possible change),[16] or laundry water from soiled diapers (G-1h)
❖ A permit is required for constructing or altering a GWS (G-3)
❖ GW is to be distributed daily (G-7)

Some Things Not Legally Required

❖ Fixtures need not be individually divertable; every fixture hooked to the system can share one diverter valve
❖ A surge tank is not required
❖ Filtration is only legally required for subsurface drip

CPC/UPC Greywater Measures Checklist

This checklist is from the California Department of Water Resources Graywater Guide, with comments from Oasis.

Drawings and Specifications (G-4)

* (G-4,a) plot plan (doesn't have to be fancy) drawn to scale showing:
* lot lines and structure
* direction and approximate slope of surface
* location of retaining walls, drainage channels, water supply lines, wells
* location of paved areas and structures
* location of sewage disposal system and 100% expansion area
* location of graywater system (Table G-1 lists required setbacks)
* number of bedrooms and plumbing fixtures
* (G-4,b) details of construction: installation, construction, and materials
* (G-4,c) log of soil formations, groundwater level, water absorption of soil
* (G-7) no irrigation point within 5' of highest known seasonal groundwater

Estimating Graywater Discharge (G-6)

* bedroom #1 (2 occupants)
* additional bedrooms (1 occupant)
* showers, tubs, wash basins: 25 gpd/occupant
* laundry: 15 gpd/occupant

Required Area (G-7)

* one irrigation zone (CPC) or three (UPC)
* each zone to distribute all graywater produced daily without surfacing
* meets Table G-2 design criteria of mini-leachfield OR
* meets Table G-2 design criteria for subsurface drip systems

Surge Tanks (G-9)

* solid, durable material, watertight when filled, protected from corrosion
* (G-5,a) anchored on dry, level, compacted soil or 3" concrete slab
* meets standards for non-potable water
* vented with locking gasketed access opening
* capacity permanently marked on tank
* "GRAYWATER IRRIGATION SYSTEM, DANGER—UNSAFE WATER" permanently marked on tank
* drain and overflow permanently connected to sewer or septic tank

Valves and Piping (G-10)

* piping downstream of water seal type trap
* piping marked "DANGER—UNSAFE WATER"
* all valves readily accessible
* backwater valves on all surge tank drain connections to sanitary drain or sewer
* (G-5,a) stub-out plumbing permanently marked

Subsurface Drip Irrigation Systems (G-11,a)

* minimum 140-mesh (115-micron) 1" filter, with a 25-gpm capacity
* filter backwash drains to the sewer system or septic tank
* emitter flow path of 1,200 microns; cv no more than 7%, flow variation no more than 10%; emitters resistant to root intrusion (see CIT list[17]). This basically requires the use of Geoflow,[11] Netafim[18] emitters, or ReWater's distribution cones.[5]
* number of emitters determined from Table G-3, minimum spacing 14"
* supply lines of PVC class 200 pipe or better and schedule 40 fittings, when pressure tested at 40 psi, drip-tight for 5 minutes
* supply lines 8" deep, feeder lines (poly or flexible PVC) 9" deep
* downstream pressure does not exceed 20 psi
* each irrigation zone has automatic flush valve and vacuum breaker

Mini-Leachfield Systems (G-11,b)

Don't do this system!

Inspection (G-5,a)

* system components identified as to manufacturer
* irrigation field installed at same location as soil test, if required
* installation conforms with approved plans

Testing (G-5,b)

* surge tank remains watertight as tank is filled with water
* flow test shows all lines and components remain watertight

Annotated CPC/UPC Greywater Law

This is the full text of Appendix G of the 2000 CPC (California Plumbing Code), with notations on the slight but devastating differences in the 2000 UPC (Uniform Plumbing Code) greywater code noted. The UPC is the model code used by 22 western states and some eastern ones.

Since the CPC has been revised more recently, you could argue that it more accurately represents the state of the art in greywater regulation than some of the anachronistic, burdensome requirements still in the UPC.

We've provided extensive annotations to make the implications of the code for builders more understandable.

*Code buffs will note a striking similarity between Appendix G (greywater) and UPC Appendix K (septic systems). Though the hardware and issues are completely different, these greywater laws started out in life as crude adaptations of tradition-steeped septic tank regulations. This helps account for how poorly they fit greywater reality. The effect is that the UPC/CPC are larded with every possible obstacle to greywater use and plenty of false leads. We suggest some revisions following the code, but really **this whole law would best be scrapped and replaced with something along the lines of the Arizona code.***

APPENDIX G GRAYWATER SYSTEMS
Title 24, Part 5, California Administrative Code

G-1 Graywater Systems. (General)

(a) The provisions of this Appendix shall apply to the construction, installation and repair of graywater systems for <u>subsurface landscape irrigation.</u> ~~Installations shall be allowed only in single family dwellings.~~ The greywater system shall not be <u>connected to any potable water</u> system without an air gap (a space or other physical device which prevents backflow) and <u>shall not result in any surfacing</u> of the graywater. <u>Except as otherwise provided for in this Appendix, the provisions of the Uniform Plumbing Code (UPC) shall be applicable to graywater installations.</u>

Thankfully, this limitation was removed from the CPC in the last revision. It remains in the UPC.

(b) The type of system shall be determined on the basis of location, soil type, and ground water level and shall be designed to accept all graywater connected to the system from the building. The system shall discharge into subsurface irrigation fields and <u>may</u> include surge tank(s) and appurtenances, as required by the Administrative Authority.

(accessories)

A surge tank is not required

(c) <u>No graywater system, or part thereof, shall be located on any lot other than the lot which is the site of the building or structure which discharges the graywater;</u> nor shall any graywater system or part thereof be located at any point having less than the <u>minimum distances indicated in Table G-1.</u>

Important! Be sure to check these distances early on in the design.

See G-4a

(d) No permit for any graywater system shall be issued until a <u>plot plan with appropriate data</u> satisfactory to the Administrative Authority has been submitted and approved. When there is <u>insufficient lot area or inappropriate soil conditions</u> for adequate absorption of the graywater, as determined by the Administrative Authority, no graywater system shall be permitted. The Administrative Authority is a city or county.

(e) No permit shall be issued for a graywater system which would adversely impact a <u>geologically sensitive area,</u> as determined by the Administrative Authority.

i.e., help cause a landslide.

(f) Private <u>sewage disposal systems existing or to be constructed on the premises shall comply</u> with Appendix I of this code or applicable local ordinance. When abandoning underground tanks, Section 722.0 of the UPC shall apply. Also, appropriate clearances from graywater systems shall be maintained as provided in Table G-1. The <u>capacity of the private sewage disposal system, including required future areas, shall not be decreased</u> by the existence or proposed installation of a graywater system servicing the premises.

In other words, even though you can now build a greywater system, you still have to build (and pay for) a complete conventional treatment system, however redundant, unnecessary, or environmentally damaging it may be. Hopefully some flexibility will be added in a future revision. Utah, for example, figures that if you go to the trouble of making a year-round greywater system, you should be allowed a smaller septic system.

(g) <u>Installers of graywater systems shall provide an operation and maintenance manual,</u> acceptable to the Administrative Authority, to the owner of each system. Graywater systems require regular or periodic maintenance.

(h) The Administrative Authority shall provide the applicant a copy of this Appendix.

The omission of kitchen sinks and dishwashers is unnecessary and unfortunate; this high solids water is a hardware design problem, not a public health or horticultural problem. The laundry water provision is obviously unenforceable; the practical effect is to require that there be an easy means of diverting laundry water to the septic/sewer at will, which is a good idea and required elsewhere anyway.

G-2 Definitions.

Graywater is untreated waste water which has not come into contact with toilet waste. Graywater includes waste water from bathtubs, showers, bathroom wash basins, clothes washing machines, and laundry tubs, or an equivalent discharge as approved by the Administrative Authority. It does not include waste water from kitchen sinks, photo lab sinks, dishwashers, or laundry water from soiled diapers.

Assure the Administrative Authority (AA) you would never send diaper water to a GWS.

Surfacing of graywater means the ponding, running off, or other release of graywater from the land surface.

This definition would seem to allow free flow Branched Drain outlets where the greywater is exposed for just a moment before disappearing sub-mulch in a contained basin (but see G-13c).

G-3 Permit.

It shall be unlawful for any person to construct, install or alter, or cause to be constructed, installed or altered any graywater system in a building or on a premises without first obtaining a permit to do such work from the Administrative Authority.

G-4 Drawings and Specifications.

The Administrative Authority may require any or all of the following information to be included with or in the plot plan before a permit is issued for a graywater system:

The plot plan is not required to be fancy. However, a fancy CAD drawing definitely will put the AA more at ease by making it look more like you know what you're doing.

(a) Plot plan drawn to scale completely dimensioned, showing lot lines and structures, direction and approximate slope of surface, location of all present or proposed retaining walls, drainage channels, water supply lines, wells, paved areas and structures on the plot, number of bedrooms and plumbing fixtures in each structure, location of private sewage disposal system and 100 percent expansion area or building sewer connecting to public sewer, and location of the proposed graywater system.

(b) Details of construction necessary to ensure compliance with the requirements of this Appendix together with full description of the complete installation including installation methods, construction and materials as required by the Administrative Authority.

Greywater systems are forgiving because the water is distributed over such a wide area (see also Permit Example 2). Table G-2 is plenty accurate. Go this route if you possibly can. A perk test can simply be digging a hole to the depth of the GW discharge, filling it with water, and timing the minutes per inch of percolation. Even if you have a bore log for a septic tank, foundation, or well, it may unnecessarily complicate the process as compared to just using Table G-2.

(c) A log of soil formations and ground water level as determined by test holes dug in close proximity to any proposed irrigation area, together with a statement of water absorption characteristics of the soil at the proposed site as determined by approved percolation tests. In lieu of percolation tests, the Administrative Authority may allow the use of Table G-2, an infiltration rate designated by the Administrative Authority, or an infiltration rate determined by a test approved by the Administrative Authority.

(d) A characterization of the graywater for commercial, industrial, or institutional systems, based on existing records or testing.

G-5 Inspection and Testing.

This provision (which, unfortunately, is not in the UPC) is a godsend; if there is nothing currently permitted that makes sense for your site (strong likelihood), you can install a permitted stub-out (especially important for slab construction). It just needs to conform to UPC and be labeled. Another great option is to plumb black- and greywater completely separately with a spot in mind for splicing in a diverter valve in the future. Separate plumbing WITHOUT GW stub-outs is easily done and completely legal now under ALL plumbing codes.

(a) Inspection

General inspection provisions

1. All applicable provisions of this Appendix and of Section 103.5 of the UPC shall be complied with.
2. System components shall be properly identified as to manufacturer. *Except common plumbing parts*
3. Surge tanks shall be installed on dry, level, well-compacted soil if in a drywell, or on a level, three inch concrete slab or equivalent, if above ground. *Underground tanks should be anchored against floating.*
4. Surge tanks shall be anchored against overturning.
5. If the irrigation design is predicated on soil tests, the irrigation field shall be installed at the same location and depth as the tested area.
6. Installation shall conform with the equipment and installation methods identified in the approved plans. *Another reason to use Table G-2 instead of tests*
7. Graywater stub-out plumbing may be allowed for future connection prior to the installation of irrigation lines and landscaping. Stub-out shall be permanently marked "GRAYWATER STUB-OUT, DANGER - UNSAFE WATER."

(b) Testing

1. Surge tanks shall be filled with water to the overflow line prior to and during inspection. All <u>seams and joints shall be left exposed and the tank shall remain watertight</u>.
2. A flow test shall be performed through the system to the point of graywater irrigation. <u>All lines and components shall be watertight.</u>

At first glance it may seem ridiculous to tightly seal pipes when the end objective is to dribble the water into the groun[d] anyway. However, th[is] may prevent damag[e] ing root encroachment.

G-6 Procedure for Estimating Graywater Discharge.

(a) Single Family Dwellings and Multi-Family Dwellings

These numbers are pretty reasonable. It is simplest to go with them, unless you generate quite a bit more or less greywater and you want to size the system accordingly. S[ee] also Permit Example 2 in Appendix F, Greywater Calcs.

The Administrative Authority may utilize the <u>graywater discharge procedure</u> listed below, water use records, or calculations of local daily per person interior water use:

1. The number of occupants of each dwelling unit shall be calculated as follows:

First Bedroom	2 occupants
Each additional bedroom	1 occupant

2. The estimated graywater flows of each occupant shall be calculated as follows:

Showers, bathtubs and wash basins	25 GPD/occupant
Laundry	15 GPD/occupant

3. The total number of occupants shall be multiplied by the applicable estimated graywater discharge as provided above and the type of fixtures connected to the graywater system.

(b) Commercial, Industrial, and Institutional Projects

The Administrative Authority may utilize the graywater discharge procedure listed below, water use records, or other documentation to estimate graywater discharge:

1. The square footage of the building divided by the occupant load factor from UPC Table 10-A equals the numbers of occupants.

2. The number of occupants times the flow rate per person (minus toilet water and other disallowed sources) from UPC Table I-2 equals the estimated graywater discharge per day.

The graywater system shall be designed to distribute the total amount of estimated graywater discharged daily.

G-7 Required Area of Subsurface Irrigation.

Each irrigation zone shall have a <u>minimum effective irrigation area</u> for the type of soil and infiltration rate to distribute all graywater produced daily, pursuant to Section G-6, without surfacing. The required irrigation area shall be <u>based on the estimated graywater discharge, pursuant</u> to Section G-6, <u>size of surge tank,</u> or a <u>method determined by the Administrative Authority.</u> ~~Each proposed graywater system shall include at least two irrigation zones and each irrigation zone shall be in compliance with the provisions of this Section.~~

An overly large dosing surge tank could deliver more water at once than the system could handle if the water from the fixtures just went straight through.

If a mini-leachfield irrigation system is used, the required square footage shall be determined from Table G-2, or equivalent, for the type of soil found in the excavation. The area of the irrigation field shall be equal to the aggregate length of the perforated pipe sections within the irrigation zone times the width of the proposed mini-leachfield trench.

<u>No irrigation point shall be within five vertical feet of the highest known seasonal</u> <u>groundwater</u> <u>nor where graywater may contaminate the groundwater or ocean</u> water. The applicant shall supply evidence of ground water depth to the satisfaction of the Administrative Authority.

Flexibility!

Thankfully, this has been removed from the CPC. The UPC requires three. (See needed improvements, following.)

The AA will generally know this for a given area; if not, depth to water in nearby wells may suffice. Greywater systems are better for high groundwater; see Greywater Purification.

G-8 Determination of Irrigation Capacity.

(a) In order to determine the absorption quantities of <u>soils other than those listed in Table</u> <u>G-2</u>, the proposed site may be subjected to percolation tests acceptable to the Administrative Authority or determined by the Administrative Authority.

Greywater systems
can function quite
well outside of these
boundaries, especially
with high perk, so
hopefully the AA will
allow some latitude.

(b) When a percolation test is required, no mini-leach field system or subsurface drip irrigation system shall be permitted if the test shows the absorption capacity of the soil is
<u>less than 60 minutes/inch or more rapid than 5 minutes/inch</u>, unless otherwise permit-
ted by the Administrative Authority.

(c) The irrigation field size may be computed from Table G-2, or determined by the Administrative Authority or a designee of the Administrative Authority.

G-9 Surge Tank Construction. (FIG. 1) *Note: A surge tank is not required.*

Structural calcula-
tions would be overkill
for the ≤100 gal
surge tanks employed
in most systems, so
hopefully the AA will
not require them.

(a) <u>Plans for surge tanks</u> shall be submitted to the Administrative Authority for approval.
The plans shall show the data required by the Administrative Authority and may include dimensions, <u>structural calculations</u>, and bracing details.

Plastic or fiberglass work well, steel drums last only a few years.

(b) Surge tanks shall be constructed of solid, durable materials, <u>not subject to excessive
corrosion</u> or decay and shall be watertight.

(c) Surge tanks shall be vented as required by Chapter 9 of this Code and shall have a
<u>locking, gasketed access opening</u>, or approved equivalent, to allow for inspection and
cleaning.

Thankfully the AA has
eliminated the drain
as a requirement here
(though it remains in
the UPC). A drain facili-
tates cleaning of above-
ground tanks. Caution:
There is rarely enough
fall to feed from fixtures
and run overflow and
drain from a surge tank
to the septic/sewer by
gravity. Every foot away
from the line connecting
fixtures to septic/sewer
costs ½" of fall; ¼" per
foot coming in and go-
ing out. A wide, shallow
surge tank helps con-
serve fall.

(d) Surge tanks shall have the <u>rated capacity permanently marked on the unit. In addition,</u>
"GRAYWATER IRRIGATION SYSTEM, DANGER - UNSAFE WATER" shall be perma-
nently marked on the surge tank.

(e) Surge tanks installed above ground shall have an ~~drain and~~ <u>overflow,</u> separate from
the line connecting the tank with the irrigation fields. The <u>overflow shall have a per-
manent connection to a sewer or to a septic tank</u>, and shall be protected against sewer
line backflow by a <u>backwater valve</u>. The <u>overflow shall not be equipped with a shut-off</u>
valve.

(f) The <u>overflow and drain pipes shall not be less in diameter than the inlet pipe.</u> The <u>vent
size shall be based on the total graywater fixture units</u>, as outlined in UPC Table 7-5
or local equivalent. <u>Unions</u> or equally effective fittings shall be provided for all piping
connected to the surge tank."

(g) Surge tanks shall be structurally designed to withstand anticipated loads. <u>Surge tank
covers shall be capable of supporting an earth load of not less than 300 pounds per
square foot</u> when the tank is designed for underground installation.

The law should
require that under-
ground tanks be
anchored against
popping to the sur-
face. When they are
empty and the soil
is saturated, the
upward force is equal
to the weight of the
water or mud they
displace: at least
450 lbs for a 55 gal
tank!

(h) Surge tanks may be installed below ground in a dry well on compacted soil, or buried
if the tank design is approved by the Administrative Authority. The system shall be
designed so that the <u>tank overflow will gravity drain to a sanitary sewer line or septic
tank.</u> The tank must be protected against sewer line backflow by a <u>backwater valve.</u>

(i) Materials

1. Surge tanks shall meet nationally recognized standards for non potable water and
shall be approved by the Administrative Authority.

2. <u>Steel surge tanks shall be protected from corrosion,</u> both externally and internally,
by an approved coating or by other acceptable means.

G-10 Valves and Piping. (FIG. 1)

An inexpensive swing
check valve usually suffic-
es for this requirement.
You could also point out
that conventional plumb-
ing allows sewage to
back out of a shower, for
example, and no back-
water valve is required
to prevent this. Perhaps
because it is so out of
line, this requirement is
rarely enforced.

Graywater piping discharging into a surge tank or having a direct connection to a sanitary drain or sewer <u>piping shall be downstream of an approved waterseal type trap(s)</u>. If no
such trap(s) exists, an approved vented running trap shall be installed upstream of the connection to protect the building from any possible waste or sewer gasses. Vents and venting shall
meet the requirements in Chapter 9 of the UPC. <u>All graywater piping shall be marked</u> or shall
have a continuous tape marked with the words "DANGER - UNSAFE WATER." All <u>valves, including the three-way valve, shall be readily accessible</u> and shall be approved by the Administrative Authority. A <u>backwater valve</u>, installed pursuant to this Appendix, shall be provided on
all surge tank drain connections to the sanitary drain or sewer piping.

There are few systems proven to meet these requirements—see "Subsurface Drip Irrigation," p. 7.

G-11 Irrigation Field Construction.

The Administrative Authority may permit <u>subsurface drip</u> irrigation, <u>mini-leach field</u> or <u>other equivalent irrigation methods which discharge graywater in a manner which ensures that the graywater does not surface</u>. Design Standards for subsurface drip irrigation systems and mini-leach field irrigation systems follow:

This key provision gives the AA latitude to permit any system (see "surface" definition G-2). "Equivalency" with subsurface drip or Mini-Leachfields could be very broad or narrowly interpreted.

(a)   Standards for a subsurface drip irrigation system are:

Measured when the filter is new or freshly cleaned.

1.   Minimum 140 mesh (115 micron) <u>filter with a capacity of 25 gallons per minute</u>, or equivalent, filtration, sized appropriately to maintain the filtration rate shall be used. The filter <u>back-wash and flush discharge shall be caught, contained and disposed of to the sewer system, septic tank, or with approval of the Administrative Authority, a separate mini-leach field</u> sized to accept all the back-wash and flush discharge water. Filter <u>back wash water and flush water shall not be used for any purpose</u>. Sanitary procedures shall be followed when handling filter back-wash and flush discharge of graywater.

There is no practical reason not to use it for irrigation, if the hardware can handle it.

Doing all this research is not necessary. This boils down to a requirement to use GeoFlow[11] or Netafim brand underground drip tubing, or ReWater[5] distribution cone emitters.

2.   Emitters shall have minimum flow path of 1200 microns and shall have a coefficient of manufacturing variation (Cv) of no more than seven percent. <u>Irrigation system design shall be such that emitter flow variation shall not exceed plus or minus ten percent.</u> Emitters shall be recommended by the manufacturer for subsurface use and graywater use, and shall have demonstrated resistance to root intrusion. For emitter ratings refer to: Irrigation Equipment Performance Report, Drip Emitters and Micro-Sprinklers, Center for Irrigation Technology, California State University, 5730 N. Chestnut Avenue, Fresno, California 93740-0018.

Hopefully no one will actually think to enforce this requirement, which would be to require unnecessary extra hardware on slopes and prohibit long runs of emitters. With 7% manufacturing variation, 3% pressure differential would take the flow to the variability limit. The goal is to preclude high flow which could result in per emitter ponding. The means of achieving that goal should be up to the installer.

3.   <u>Each irrigation zone shall be designed to include no less than the number of emitters specified in Table G-3</u>, or through a procedure designated by the Administrative Authority. <u>Minimum spacing between emitters is 14 inches</u> in any direction.

4.   The system design shall provide user controls, such as valves, switches, timers, and other controllers as appropriate, to <u>rotate the distribution of graywater between irrigation zones</u>.

These burial requirements are unnecessarily deep for most conditions. According to the Department of Water Resources the literal requirement is that the line be inches below "the surface" and since the type of surface is not specified, it could be mulch instead of soil. This would allow for future inspection and service, as well as changing the system as the landscaping evolves.

5.   All drip irrigation <u>supply lines shall be</u> polyethylene tubing or <u>PVC class 200 pipe</u> or better and <u>schedule 40 fittings</u>. All joints shall be properly solvent-cemented, inspected and pressure tested at 40 psi, and shown to be <u>drip tight for five minutes, before burial</u>. All supply lines will be <u>buried at least eight inches deep</u>. <u>Drip feeder lines can be poly or flexible PVC tubing and shall be covered to a minimum depth of nine inches</u>.

6.   Where pressure at the discharge side of the pump exceeds 20 pounds per square inch (psi), a <u>pressure reducing valve</u> able to maintain downstream pressure no greater than 20 psi shall be installed downstream from the pump and before any emission device.

It's hard to imagine a more pointless waste of electricity; instead use a pump which does not produce as much pressure. In the event that the emitters are significantly higher in elevation than the pump, hopefully the AA would focus on the pressure at the emitters, rather than the higher pressure at the pump.

Flush valve at end, anti-siphon at beginning

7.   Each irrigation zone <u>shall include a flush valve/anti-siphon valve</u> to prevent back siphonage of water and soil.

(b)   Standards for the mini-leach field system are:

Same as a septic tank leachfield! (Details in UPC or local statute.)

1.   Perforated sections shall be a <u>minimum 3-inch diameter</u> and shall be constructed of perforated high density polyethylene pipe, perforated ABS pipe, perforated PVC pipe, or other approved materials, provided that sufficient openings are available for distribution of the graywater in the trench area. Material, construction and perforation of the piping shall be in compliance with the appropriate <u>absorption field drainage piping standards</u> and shall be approved by the Administrative Authority.

Too bad; 1" would be better for some designs. It will be all but impossible to get even distribution along the length of a pipe which takes so much water just to...

2. Clean stone, gravel, or similar filter material acceptable to the Administrative Authority, and varying in size between 3/4 inch to 2 inches shall be placed in the trench to the depth and grade required by this Section. Perforated sections shall be laid on the filter material in an approved manner. The perforated sections shall then be covered with filter material to the minimum depth required by this Section. The filter material shall then be covered with landscape filter fabric or similar porous material to prevent closure of voids with earth backfill. No earth backfill shall be placed over the filter material cover until after inspections and acceptance.

3. Irrigation fields shall be constructed as follows:

	Minimum	Maximum
Number of drain lines per irrigation zone	1	—
Length of each perforated line	—	100 feet
Bottom width of trench	6 inches	18 inches
Total depth of trench	17 inches	18 inches
Spacing of lines, center to center	4 feet	—
Depth of earth cover of lines	9 inches	—
Depth of filter material cover of lines	2 inches	—
Depth of filter material beneath lines	3 inches	—
Grade of perforated lines	level	3 inches / 100 feet

Precision shovel work! See Figure 2, p. 10.

Precision leveling

G-12 Special Provisions.

(a) Other collection and distribution systems may be approved by the Administrative Authority as allowed by Section 301 of the UPC.

(b) Nothing contained in this Appendix shall be construed to prevent the Administrative Authority from requiring compliance with stricter requirements than those contained herein, where such stricter requirements are essential in maintaining safe and sanitary conditions or from prohibiting graywater systems. The prohibition of graywater systems or more restrictive standards may be adopted by the Administrative Authority by ordinance after a public hearing.

G-13 Health and Safety.

(a) Graywater may contain fecal matter as a result of bathing and/or washing of diapers and undergarments. Water containing fecal matter, if swallowed, can cause illness in a susceptible person. Therefore, graywater shall be not be contacted by humans, except as required to maintain the graywater treatment and distribution system.

(b) Graywater shall not include laundry water from soiled diapers.

(c) Graywater shall not be applied above the land surface or allowed to surface and shall not be discharge directly into or reach any storm sewer system or any water of the United States.

(d) Graywater shall not be used for vegetable gardens.

Table G-1 Location of Graywater System.

Minimum Horizontal Distance (in feet) From	Surge Tank	Irrigation Field	Tank	Field
Buildings or structures[1]	5 ft[2]	8 ft[3]	0 ft	2–8 ft
Property line adjoining private property	5 ft[4]	5 ft	0 ft	3 ft
Water supply wells[5]	50 ft	100 ft	2 ft	50 ft
Streams and lakes[5]	50 ft	50 ft	25–50 ft	25–50 ft
Seepage pits or cesspools	5 ft	5 ft	0 ft	0–5 ft
Disposal field & 100% expansion area	5 ft	4 ft[6]	0 ft	0 ft
Septic tank	0 ft	5 ft[7]	0 ft	0 ft
On-site domestic water service line	5 ft	5 ft[8]	5 ft	5 ft
Pressure public water main	10 ft	10 ft[9]	10 ft	10 ft
Water ditches	50 ft	50 ft	25–50 ft	25–50 ft

Notes: When mini-leach fields are installed in sloping ground, the <u>minimum horizontal distance between any part of the distribution system and ground surface shall be fifteen feet.</u>

This relic of the septic code effectively prohibits installations on slopes greater than 1:15, or 6.67%, unless you terrace the ground. Hopefully your AA will not catch this, since on some soils far steeper, unterraced slopes could be irrigated without surfacing.

1. Including porches and steps, whether covered or uncovered, but does not include car ports, covered walks, driveways and similar structures.
2. The distance <u>may be reduced to zero feet</u> for above ground tanks if approved by the Administrative Authority.
3. The distance <u>may be reduced to two feet</u>.
4. For subsurface drip irrigation systems, 2 feet from property line.
5. Where special hazards are involved, the <u>distance may be increased</u> by the Administrative Authority.
6. Applies to the <u>mini-leach field type system only</u>. Plus two feet for additional foot of depth in excess of one foot below the bottom of the drain line.
7. Applies to <u>mini-leach field type system only</u>.
8. A <u>two foot separation is required for subsurface drip systems</u>.
9. For parallel construction or for crossings, approval by the Administrative Authority shall be required.

Table G-2 Mini-Leach Field Design Criteria of Six Typical Soils.

Type of Soil	Minimum sq. ft. of irrigation area per 100 gallons of estimated graywater discharge per day				Maximum absorption capacity, minutes per inch, of irrigation area for a 24-hour period.
1. Coarse sand or gravel	20	40	20	13	5
2. Fine sand	25	50	25	17	12
3. Sandy loam	40	80	40	27	18
4. Sandy clay	60	120	60	40	24
5. Clay with considerable sand or gravel	90	180	90	60	48
6. Clay with small amount of sand or gravel	120	240	120	80	60
		6"	12"	18"	

Equates to this many linear feet for these 3 trench widths

Table G-3 Subsurface Drip Design Criteria of Six Typical Soils.

Type of Soil	Maximum emitter discharge (gal/day)	Minimum number of emitters per gpd of graywater production
1. Sand	1.8	.6
2. Sandy loam	1.4	.7
3. Loam	1.2	.9
4. Clay loam	.9	1.1
5. Silty clay	.6	1.6
6. Clay	.5	2.0

<u>Use the daily graywater flow calculated in Section G-6 to determine the number of emitters per line.</u>

This means the number of emitters per irrigation zone. For example, if you have 100 gpd of greywater according to Table G-6, and clay soil, you need two emitters per gpd of greywater generation, or 200 per zone. With emitters 14" apart, that's 234' of line. While the CPC has been improved to allow only one zone, the UPC requires three (remember, this is in addition to a septic or sewer connection).

CPC Appendix G Figure 1: Graywater System (conceptual)

Caution: These figures are purely conceptual and can be misleading, especially to inspectors. They may, unfortunately, be the only information your inspector has seen on greywater systems. This drawing should have a note that running traps are generally illegal, and they are required for GW surge tanks only if the fixtures lack traps (this is in the text of the law).

Notes on the UPC figures (not shown). The CPC had these figures before, but they were thrown out in the last revision. The UPC figures are worse. They show sewage ejector pump tanks built to withstand pressure in the source line, which may build up if the pump fails. Since greywater tanks have overflows, this type of tank is not necessary. Both cleanouts and backwater valves should be shown above the surface of the ground or with extended access ports...

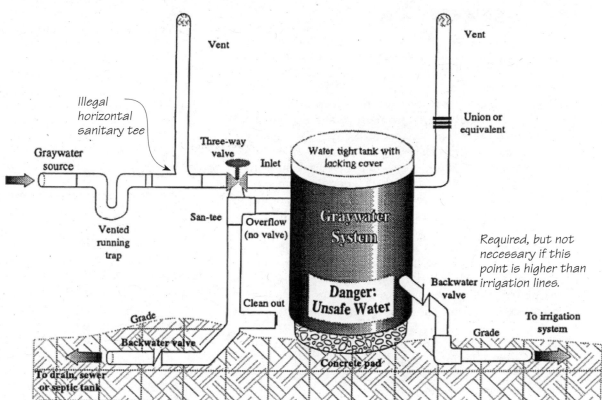

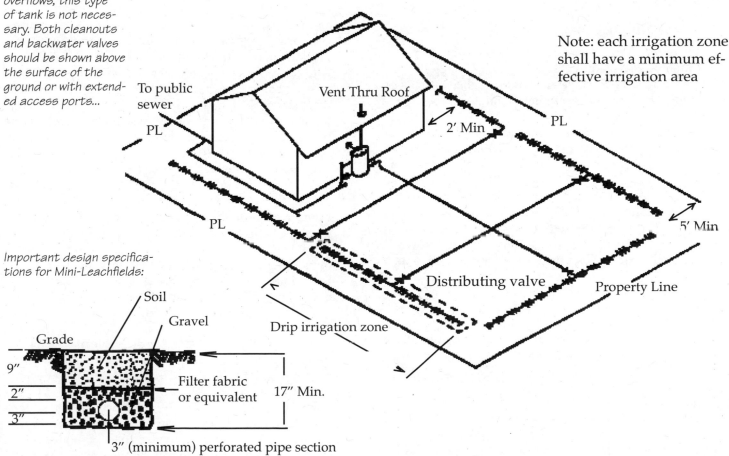

CPC Appendix G Figure 2: Irrigation Layout (conceptual)

Note: each irrigation zone shall have a minimum effective irrigation area

Important design specifications for Mini-Leachfields:

35

Needed Improvements to CPC/UPC Greywater Law

Assembly Bill 313 rectified many of the problems we'd identified with the CPC in earlier editions of this book, but the UPC still has most of these. There is still a long way to go with both laws.

These unrealistic greywater laws probably have *increased* the public health threat from greywater systems by lowering the legal compliance rate virtually to zero. Santa Barbara, for example, has issued approximately 10 permits for greywater systems since greywater use was legalized in 1989.[19] This is in an area with 200,000 people, as many as 40% of whom were using greywater in the drought of the 1990s.[20] So many requirements are obviously overkill that the entire law, including some very sensible provisions, is dismissed by the public as a source of design guidance. A more reasonable regulatory stance would lead to greater participation and a reduction in risk from the perpetuation of unregulated systems. As California's law is being taken as a model for other states and countries, this is all the more vital.

The best action would be to abandon the current CPC/UPC laws and adopt an Arizona-style tiered approach. Failing this, the incremental improvements below would be steps in the right direction.

To campaign for better laws in California, direct your comments to the agency in charge, the State Department of Water Resources[16]—and be nice. These people have worked very hard to get this law in place against considerable resistance.

General Suggestions

❖ **Wherever appropriate, require achievement of performance goals** (e.g., ecologically and biologically safe treatment of wastewater), with explicit designs as options, rather than specifying mandatory techniques to be used.

❖ **Be more realistic about the quantitative health threat from greywater systems.** There is a long history of surface greywater reuse, with systems far, far less safe than those specified in the current law, which has not produced a single documented case of greywater-transmitted illness in the United States. In Australia, greywater is legally distributed through sprinklers with 6' throw. The City of Los Angeles Greywater Pilot Project[21] showed that greywater makes a negligible contribution to the pathogens in soil, while dog feces, for example, contribute a significant amount of pathogens to the suburban environment. Even the worst illegal greywater systems don't stand out among myriad sources that besiege our bodies with pathogens in the course of ordinary life. *The actual health threat is plenty small enough to include ecological and practical considerations on equal footing with public health considerations.*

❖ **Consider exposure from required maintenance in comparing the relative health risk of systems.**

❖ **Local jurisdictions should consider the effect of high permit fees on participation in the legal process.** In our area a greywater permit costs $75, increasing the attractiveness of simple, illegal systems, which already have dramatically superior cost/benefit ratios to currently legal systems in most situations (and often cost less than $75 total!).

❖ **Change plumbing code to require greywater and blackwater to be plumbed separately for all new construction of single family homes on ¼ acre or more.** The lines should be joined after all the fixtures and vents and at or after a convenient future greywater diversion point.

Specific Suggestions

❖ G-1-a **Allow commercial and multifamily systems in the UPC** (this change has already been made in the CPC). This is a serious problem with the current law.

❖ G-1-f **Allow reduction in size or elimination of septic/sewer system if the alternative waste disposal system is capable of handling all wastes as well or better**, at the discretion of the Administrative Authority. There are sites and regions where currently mandated treatment technologies cause more ecological and health problems than proven alternatives. Regulators are allowing this in practice, and they should have clear guidelines.

❖ G-2 **Redefine kitchen sink and dishwasher effluent as "difficult-to-handle greywater" (rather than blackwater)** and allow its use at the discretion of the Administrative Authority, if the hardware is demonstrably able to handle it. This high-solids water is a (resolvable) hardware design problem, not a soil or public health problem (Branched Drains to subsoil infiltrators can handle kitchen sink water, for example—or raw sewage, for that matter).

❖ G-7 **Allow greywater systems in areas with high groundwater** at the discretion of the Administrative Authority. A proper greywater system design can provide better treatment and protect groundwater better than currently mandated systems. A specific provision requiring that a given amount of soil separate greywater from aquifers in Karst formations would be reasonable.

❖ G-7 **Eliminate from the UPC the requirement for three irrigation zones** which are each capable of accepting the entire greywater flow, if there is a disposal alternative. This ill-thought-through requirement, which has already been struck from the CPC, eliminates the possibility of meeting all the irrigation needs of an area with greywater, whether it makes sense or not. It effectively mandates the installation of a redundant freshwater irrigation system, which severely undermines the economics of some systems, particularly commercial or multifamily systems. This requirement drove the regulators' favorite manufacturer (AGWA) out of business. High-end greywater systems are capable of distributing freshwater as needed for supplemental irrigation without wasteful hardware duplication. This is a serious problem with the current law.

❖ G-7, G-8, Table G-2, Table G-3 **Explicitly allow reduction in system design loads with water-conserving fixtures.** Projects with aggressive conservation shouldn't be penalized by having to install the same size system as the worst water hogs. The current language allows local discretion in this area but the possibility is not obvious.

❖ G-8-b **Allow greywater systems across a wider range of percolation rates.** Greywater systems are safer at high percolation rates than septic systems.

❖ G-9-e **Delete the requirement for a gravity drain for surge tanks from the UPC**, as has been done with the CPC. This is a serious problem with the current law. A gravity drain is a nice convenience but it is a practical impossibility for many installations. Note that current law does not require a gravity drain for underground greywater surge tanks, septic tanks, or sewage ejector pump tanks.

❖ G-9-h **Require below-grade tanks to be anchored against popping to the surface** if conditions indicate this may be a problem. Unlike septic tanks, greywater surge tanks are often empty and experience tremendous buoyant lift under saturated soil conditions. This would protect consumers.

❖ G-11-a-2 **Modify the requirement that "system design shall be such that emitter flow variation shall not exceed plus or minus 10%"** with the phrase "in instances where greater variation could result in flows high enough to produce per emitter ponding in the soil in question."

❖ G-11-a-6 **Change wording from "pressure at pump shall not exceed 20 psi" to "pressure at any emission device shall not exceed 20 psi."** The current wording effectively precludes irrigation with adequate pressure at a location significantly higher than the pump.

❖ G-11-a-5, G-11-b-2 **Explicitly allow greywater to be distributed and emitted through lines covered by mulch** at the discretion of the Administrative Authority. This would be a great step forward.

❖ G-11-b-1 **Allow smaller diameter pipe, half-pipes in Mini-Leachfields.**

❖ Table G-1 **Allow installations on steeper slopes** where environmental conditions are such that the water will not surface.

❖ Table G-2 **Take into account the higher LTAR of mulch basins by halving the required infiltration area for systems that use them.**

❖ **Explicitly describe Branched Drain to Mulch Basins, Infiltration Beds, Leaching Chambers, and Box Troughs** (see *Create an Oasis*) as allowed system examples.

❖ Figures: **Show a greywater surge tank (usually a 55 gal drum)** rather than a sewage ejector pump tank in UPC figures. Include a note that the running trap is only required in the rare instance that the fixtures lack traps.

❖ **Eliminate the requirement for backwater valves.**

❖ **Allow greywater surge tank to be vented back through the house vents** (as is done with all septic tanks and sewers) as an alternative to a vent at the tank.

Needed Improvements to IPC Greywater Law

The IPC is the model code for most of the eastern United States. These comments are based on the 2000 IPC, Appendix C, p. 101.

General Suggestions

❖ **The regulation should lay out broad goals such as health protection and leave it at that.** This would be in keeping with the minimalist, "let the designer figure it out" philosophy of the IPC (the whole of which is less than a third the length of the UPC, with only one page on greywater). Most of the trouble with the IPC is in the form of broad prohibitions.

❖ **Starting from scratch** with Arizona-style wording would be the easiest way to accomplish this.

Specific Suggestions

❖ c101.1 **Differentiate between allowable uses for treated and untreated greywater.** As it stands, reuse for toilet flushing is allowed with disinfection only, which may not be satisfactory if BOD remains high—toilet tanks may become foul and anaerobic with stored, putrefying water. Treated greywater could be reused for other non-potable uses beside those listed, laundry for example. It is not necessary to treat greywater for irrigation in most cases.

❖ **Irrigation should be specifically allowed**, not just as an exception.

❖ c101.2 **Expand greywater definition to include all domestic wastewater other than toilet water.** Exclusion of kitchen sink water leaves this particular wastewater flow in awkward limbo in facilities with a greywater system and composting toilets. Instead, give performance requirements for dealing with high-suspended-solids, high-BOD water.

❖ c101.4 **The reservoir should be optional**, as storing greywater is not required for all system types and is generally undesirable. The "not less than 50 gal" and "not more than 72 hours retention" requirements are potentially at odds with each other. 24 hours maximum retention is a better design goal, with the tank size left up to the designer.

❖ c101.6 **Disinfection should not be required for irrigation reuse**.

❖ c101.7 **Make-up water should be optional** depending on the application. Toilet flushing requires make-up water for public health, irrigation does not.

❖ c101.8 **Overflow pipe should be the same size or greater than the influent pipe**. Allow connection to an alternate overflow, in order to allow facilities with well made, high capacity composting toilets and greywater systems without a sewer/septic hookup.
Allow septic systems to be downsized when a greywater system is safely processing most of the effluent.

Appendix E: Treatment Effectiveness

There have been very few empirical studies of the health threat from greywater. The *Greywater Pilot Project Final Report* from the Los Angeles Department of Water Reclamation is one excellent study.[21]

The excerpts from the studies that follow aid regulators and greywater system designers by shining light on the capacity of soil to purify biological contamination in two much more challenging applications: feces disposal, and land application of large volumes of partially treated wastewater, areas where research is more plentiful.

Movement of Biological and Biochemical Contamination in Soil and Groundwater

Excerpts from Excreta Disposal for Rural Areas and Small Communities reprinted with permission from the World Health Organization, Geneva, Switzerland.[22] These figures show the movement through soil and groundwater of biological and chemical contamination from pit latrines. Of special relevance to greywater system design is the finding that "when contamination does not enter the groundwater, there is practically no danger of contaminating water supplies" (text of WHO Figure 4). Bear in mind that virtually all the pathogens of concern originate in feces, and that these excerpts refer to the disposal of nearly 100% fecal matter, and that the quantity of fecal matter in greywater is very small.

WHO FIGURE 4: MOVEMENT OF
POLLUTION IN DRY SOIL

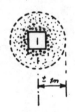

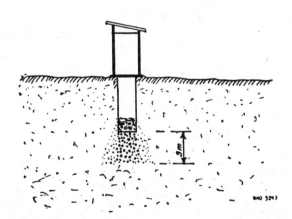

In dry soil there is relatively little migration of chemical and bacterial substances. Laterally there is practically no movement; and with excessive washing (not common in privies or septic tanks) the vertical penetration is only about 3 m (10 ft). Where the contamination does not enter the groundwater, there is practically no danger of contaminating water supplies.

Soil and Ground-Water Pollution

The study of methods of pollution of the soil and water by excreta also provides useful information concerning the design of disposal facilities, especially their location with respect to sources of drinking-water supplies. After excreta are deposited on the ground or in pits, the bacteria, unable to move much by themselves, may be transported horizontally and downward into the ground by leaching liquids or urine, or by rain water. The distance of travel of bacteria in this way varies with several factors, the most important of which is the porosity of the soil (see Fig. 4, 5, 6). Their horizontal travel through soil in this manner is usually less than 90 cm (3 ft) and the downward travel less than 3 m (10 ft) in pits open to heavy rains, and not more than 60 cm (2 ft) normally in porous soils.

Gotaas and his co-workers, studying the artificial recharge of aquifers with reclaimed sewage and other waste waters in the State of California, USA, found that bacteria were transported to a distance of up to 30 m (100 ft) from the recharge well in 33 hours, and that there was a rapid regression of bacterial count over this distance due to effective filtration and to bacterial die-off. They also found that chemical pollution travelled twice as fast. Recently, other workers, studying ground-water pollution in Alaska, noted that bacteria were traced to a distance of 15 m (50 ft) from the dosing well into which test bacteria were introduced. The width of the path of bacterial travel varied between 45 cm and 120 cm (1.5 ft and 4 ft). Regression then took place and, after a year, only the dosing well remained positive for the test organism. These investigations confirm the findings of other workers to the effect that the contamination from excreta disposal systems tends to travel downward until it reaches the water table, then moves along with the ground-water flow across a path which increases in width to a limited extent before gradual disappearance.

In the Netherlands, Baars found that, unless accompanied by a considerable amount of water, bacterial contamination did not travel more than 7.5 m (25 ft) through fine sand.

On the surface of the ground, only the earth immediately surrounding the faeces is likely to be contaminated, unless it is carried further by surface water such as rain and irrigation water, blown away by the wind, or picked up by the hair and feet of flies or other insects and animals. It has been observed in pit latrines, however, that hookworm larvae, although unable to move sideways to any appreciable extent, are likely to climb upwards along the pit walls and reach the top surface of defective wooden or earthen floors, where they lie in wait for a person with bare feet.

WHO FIGURE 5: MOVEMENT OF POLLUTION IN UNDERGROUND WATER

A = Top soil　　　　B = Water-bearing formation　　　　C = Direction of ground-water flow

WHO FIGURE 6: BACTERIAL AND CHEMICAL SOIL POLLUTION PATTERNS AND MAXIMUM MIGRATIONS

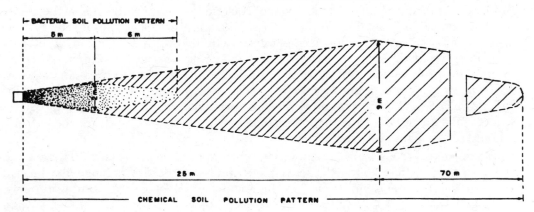

The source of contamination in these studies was human excreta placed in a hole which penetrated the ground-water table. Samples positive for coliform organisms were picked up quite soon between 4 m and 6 m (13 ft and 19 ft) from the source of contamination. The area of contamination widened out to a width of approximately 2 m (7 ft) at a point about 5 m (16 ft) from the privy and tapered off at about 11 m (36 ft). Contamination did not move "upstream" or against the direction of flow of the ground water. After a few months the soil around the privy became clogged, and positive samples could be picked up at only 2 m to 3 m (7 ft to 10 ft) from the pit. In other words, the area of soil contamination had shrunk.

The chemical pollution pattern is similar in shape to that of bacterial pollution but ex-

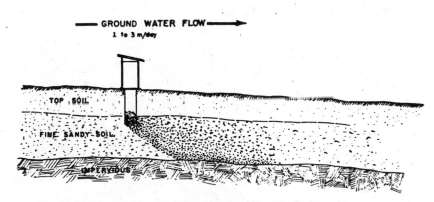

GROUND WATER FLOW
1 to 3 m/day

TOP SOIL
FINE SANDY SOIL
IMPERVIOUS

* Based on data from Caldwell & Parr and Dyer, Bhaskaran & Sekar.

tends to much greater distances. From the point of view of sanitation, the interest is in the maximum migrations and the fact that the direction of migration is always that of the flow of ground water. In locating wells, it must be remembered that the water within the circle of influence of the well flows towards the well. No part of the area of chemical or bacterial contamination may be within reach of the circle of influence of the well.

Depending upon conditions of humidity and temperature, pathogenic bacteria and ova of parasitic worms will survive varying lengths of time in the ground. Pathogenic bacteria do not usually find in the soil a suitable environment for their multiplication, and will die within a few days. On the other hand, hookworm eggs will survive as many as five months in wet, sandy soil, and three months in sewage. Hookworm disease is transmitted through contact of the skin, usually bare feet, with soil containing hookworm larvae. Other parasitic diseases are also transmitted when fresh faeces or sewage is used, during the growing season, to fertilize vegetable crops which are eaten raw.

If ground water is located near a source of infection within the distances mentioned above, it may become contaminated by harmful bacteria and by putrid chemical substances originating in the faecal decomposition. A source of infection may be some excreta deposited on the ground near by, a pit latrine, a cesspool, or a leaky sewer or sewage disposal pipe. The contaminated ground-water, which is usually shallow, may be tapped by a well used for drinking-water and other domestic purposes and may lead to further human infection and diseases such as diarrhoeas, typhoid and parathypoid fevers, cholera, and the dysenteries.

The effects of proximity of wells to latrines and the travel of faecal pollution through ground water have been investigated by various scientists. The studies of Caldwell & Parr and of Dyer, Bhaskaran & Sekar are classics which should be studied by all interested public health workers.

Location of Latrines and Other Excreta Disposal Facilities

Regarding the location of latrines with respect to sources of water supply, the following conclusions may be drawn from up-to-date information.

1. There can be no arbitrary rule governing the distance that is necessary for safety between a privy and a source of water supply. Many factors, such as slope and level of ground water and soil-permeability, affect the removal of bacteria in ground water. It is of the greatest importance to locate the privy or cesspool downhill, or at least on some level piece of land, and to avoid, if possible, placing it directly uphill from a well. Where uphill locations cannot be avoided, a distance of 15 m (50 ft) will prevent bacterial pollution of the well. Setting the privy off to either the right or the left would considerably lessen the possibility of contaminating the ground water reaching the well. In sandy soil a privy may be located as close as 7.5 m (25 ft) from a properly constructed household well if it is impossible to place it at a greater distance. In the case of a higher-yielding well, not less than 15 m (50 ft) should separate the well from a latrine.

2. In homogeneous soils the chance of ground-water pollution is virtually nil if the bottom of a latrine is more than 1.5 m (5 ft) above the ground-water table. The same may be said if the bottom of a cesspool is more than 3 m (10 ft) above the level of the ground water.

3. A careful investigation should be made before building pit privies, bored-hole latrines, cesspools, and seepage pits in areas containing fissured rocks or limestone formations, since pollution may be carried directly through solution channels and without natural filtration to distant wells or other sources of drinking-water supplies.

TABLE 3: REMOVAL RATES OF SELECTED LAND TREATMENT FACILITIES

From Green Land, Clean Streams; The Beneficial Use of Waste Water through Land Treatment.[23] Reprinted with permission from Temple University, Philadelphia, PA. This table summarizes removal rates for several facilities. A typical system distributes combined waste from a soup factory through sprinklers, after primary treatment. Loading rates in some cases are up to 6" per day. Overland flow facilities intentionally allow the water to run off, with treatment by bacteria living on the surface of the soil and plants. This presumably would be indicative of the treatment level greywater would receive if it ran off over the surface: 99%+. Most facilities operate year-round. In cases where frozen wastewater accumulates on the surface, good treatment is apparently achieved when it thaws. It is reasonable to expect that these levels would represent absolute minimums for the treatment that water would receive in residential greywater systems.

Pollutant \ Facility	Muskegon County, Michigan	Flushing Meadows, Arizona	Penn State University (woodland)	Penn State University (cropland)	Penn State University (overall)	Paris, Texas (overland flow)	Santee, California	Whittier Narrows, California	Seabrook Farms, New Jersey	Riegel Paper Co., New Jersey	Beardmore & Co., Ltd., Toronto	Shoemakers Dairies, New Jersey	Green Valley Farms, Pennsylvania
BOD	98%+	98%				99.1%	93%	99.3–99.7%	95%*	95%			
COD		100%					100%						
Total Organic Content						98.2%	75%						
Suspended Solids		100%				98.2%							
Phosphorous	98%+	87%	97%+	99%		90.0%	100%						
Nitrogen total		40–80%		100%		91.5%	100%						
Nitrogen organic	75–87%	100%	57–82%										
Nitrogen ammonia	97%	98.2%					100%						
Fluoride		50%											
Chloride					****								
Salt		Slight increase			****								
Potassium			82.8%	118%									
Magnesium			66.7%	11%									
Sodium				0.4%									
Calcium			51.9%	19%									
Boron			67.6%										
ABS (detergent)			(MBAS) 81%	98%	****		100%	97%					
Coliform total	100%						99%+	100%					
Coliform fecal		100%											
Virus pathogenic	100%						100%						
pH	**												**
Other	*****											***	***

*   Approximate measures
**   Effluent will meet USPH Drinking Water Standard
***   Effluent meets state and/or local health authority test requirement
**** There is no increase in the natural concentration in the groundwater
***** Heavy metal concentrations will be below the threshold level for fish, wildlife and agriculture

FIGURE 4: HOW TREATMENT CAPACITY AND CONTAMINATION PLUMES CHANGE
with LOCATION OF WASTEWATER APPLICATION

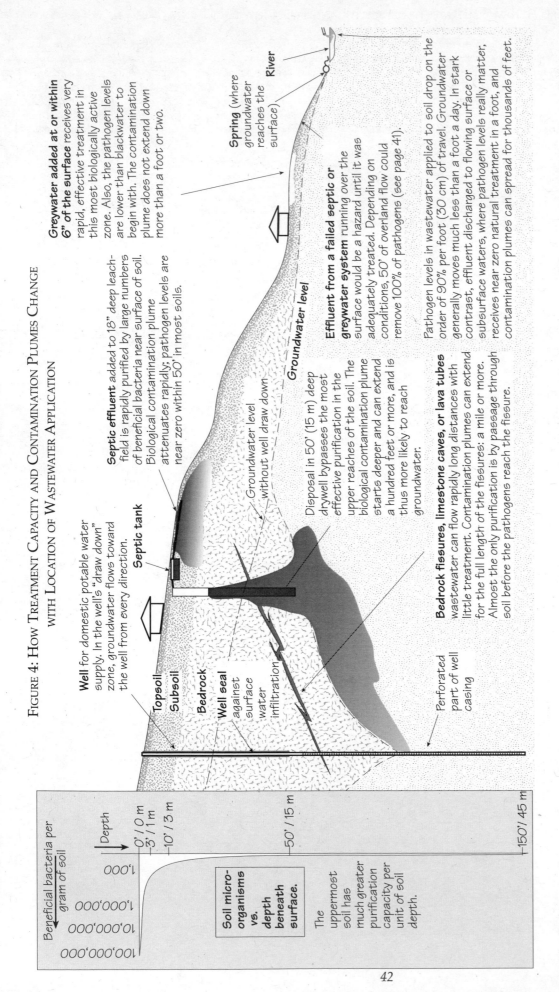

Plume extents do not equal recommended separation distances. If a septic leachfield plume generally extends 50', the separation distance should be a multiple of that to allow a safety factor. The UPC requires 100' of separation from a well to a leachfield, for example—a 2x safety factor.

The Code is not especially rational. It only requires 150' for a drywell, which provides little or no safety factor. The code separation from a greywater outlet to a well is 100', about a 10x safety factor, but only 50' from natural surface waters, about a 3x safety factor (pathogens can travel farther over the surface than through the soil).

The optimum application point is a balance between various considerations. You want to apply wastewater high to get the best treatment, safeguard groundwater, and improve reuse, but not so high that it daylights.

Though excellent treatment can be attained on the surface (see Overland Flow, p. 43), wastewater may be contacted by children or animals, or run off into surface water before it is treated.

In bulk flowing water, there is almost no natural treatment other than dilution, the treatment modality considered least desirable by the World Health Organization. Contamination can spread for miles downstream. This is a major weakness of current wastewater treatment thinking in the US, which favors a pipe dumping artificially treated effluent directly into natural waters. If there is any problem — the plant runs out of chlorine, the power goes off—or for that matter, even if there is not, the waste is injected into the worst place for it in the entire water cycle. Bulk water flows have thousands of times less purification capacity than water in soil, and they are the point in

Appendix F: Sample Permit Submissions

Example 1: Town—City of Santa Barbara (Branched Drain)

All versions of the Branched Drain system meet the Arizona and New Mexico requirements. If you use one of the subsurface versions of the Branched Drain network, it is possible to get a permit for it under the CPC/UPC.

This system is one of only a handful permitted in Santa Barbara in 10 years under the CPC. The site and hardware are described in the system examples appendix of *Create an Oasis*.

Our Mulch Basins were force-fit to the Mini-Leachfield model in the law. If I were to do it again I'd go for the "other equivalent system" option and let them have more natural shape and dimensions, which is what we ended up doing anyway.

FIGURE 5: GREYWATER CALCULATIONS
FOR PERMIT EXAMPLE 1

1st bedroom	2
2nd bedroom	1
Total	3

Greywater generation per occupant (gpd = gallons per day)

Gpd/occupant (showers, bathtub, & wash basins)	25
Gpd/occupant (laundry)	15
Total	40

Greywater source: 40 gpd/occupant x 3 occupants = 120 gpd (840 gallons per week)

Irrigation:

This is twice the actual greywater generation. Seize any opportunity to get the greywater numbers down. This will help counteract the quadruple redundancy required elsewhere in the law.

Soil type: sandy loam (as per infiltration test)

Grade: 1/4" per foot

Irrigation area required: 48 square feet
  Equals 32 linear feet of 18" wide trench per zone

Specifications [as per California Uniform Plumbing Code (UPC), Appendix J]

•Mini-leach fields are in two zones each with a 48 square foot area
  [i.e. each capable of holding the daily greywater flow].

•All exposed greywater plumbing will be labeled

   "Greywater irrigation system: Danger - unsafe water"

•Fittings shall be Schedule 40 2" or 1 1/2" PVC or ABS

•**Note:** Surge tank NOT required by UPC Appendix J - Section J-1(b) [page 1]

   "The system .... MAY include surge tank(s)......

   [see also Administrative Authority for approval of OTHER collection and distribution systems in Appendix J, Section J-12(a)]

CITY OF SANTA BARBARA
DIVISION OF LAND USE CONTROL
BUILDING AND SAFETY SECTION
A P P R O V E D
AUG 08 1996

By_____
The stamping of this set of plans and specifications SHALL NOT be held to permit or to be an approval of the violation of any provisions of any city ordinance or state law. It is unlawful to make any changes or alterations to same without written permission from the Division of Land Use Control, Building and Safety Section, City of Santa Barbara.
Approved by Building

SBPPC

FIGURE 7: MULCH BASIN DETAIL FOR PERMIT

At one point they indicated wood chips would not be acceptable but they ended up going for it, thank goodness.

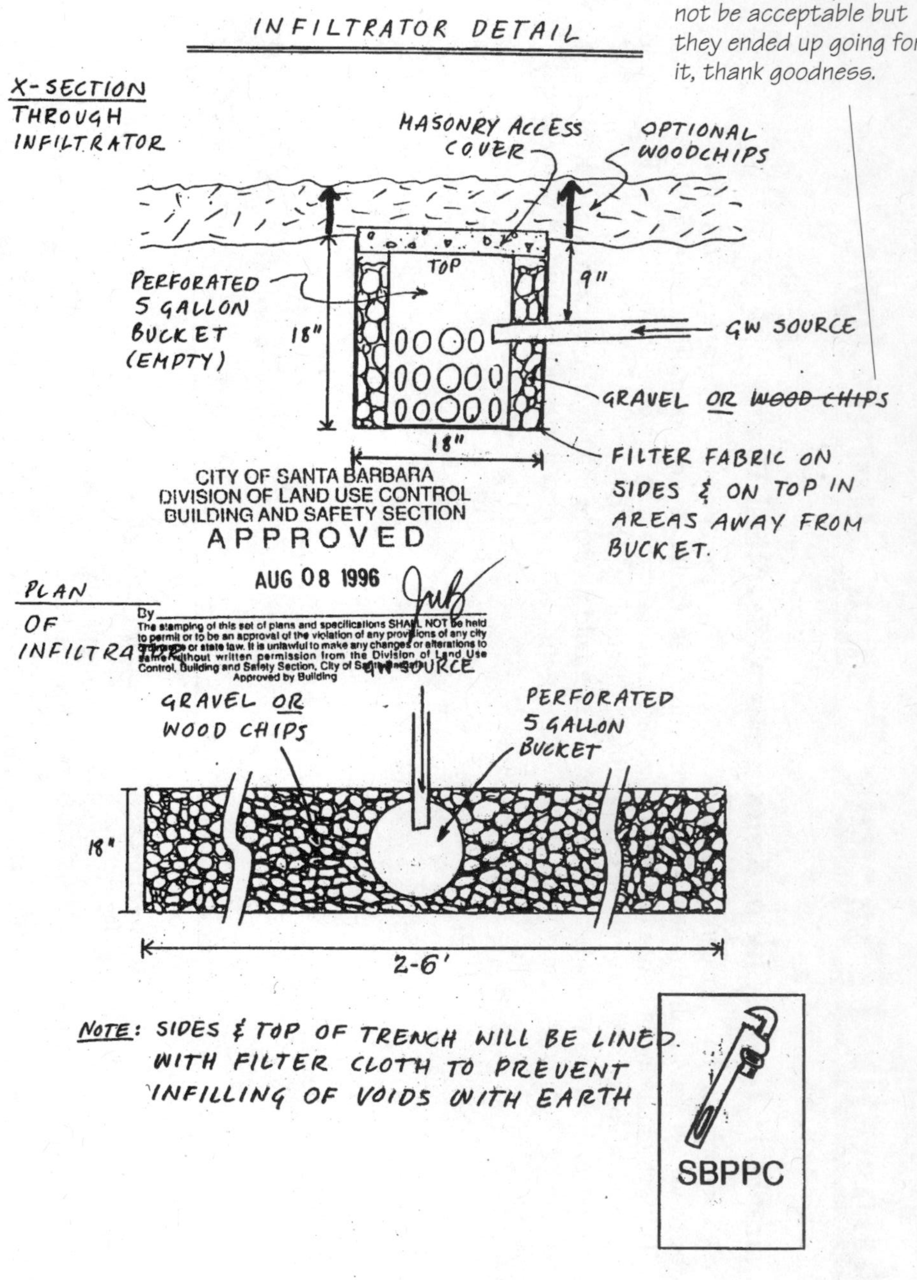

OPTIONAL WOODCHIPS

GW SOURCE

GRAVEL OR WOODCHIPS

FILTER FABRIC ON SIDES & ON TOP IN AREAS AWAY FROM BUCKET.

INFILTRATOR DETAIL

MASONRY ACCESS COVER

TOP

9"

18"

18"

X-SECTION THROUGH INFILTRATOR

PERFORATED 5 GALLON BUCKET (EMPTY)

CITY OF SANTA BARBARA
DIVISION OF LAND USE CONTROL
BUILDING AND SAFETY SECTION
APPROVED
AUG 08 1996

The stamping of this set of plans and specifications SHALL NOT be held to permit or to be an approval of the violation of any provisions of any city ordinance or state law. It is unlawful to make any change or alteration to these plans without written permission from the Division of Land Use Control, Building and Safety Section, City of Santa Barbara.
Approved by Building

PLAN OF INFILTRATOR

GW SOURCE

PERFORATED 5 GALLON BUCKET

GRAVEL OR WOOD CHIPS

18"

2'-6'

NOTE: SIDES & TOP OF TRENCH WILL BE LINED WITH FILTER CLOTH TO PREVENT INFILLING OF VOIDS WITH EARTH

SBPPC

FIGURE 6: PLOT PLAN FOR PERMIT

Dipper box is here under a deck, with fairly easy access. Sewer diversion is just before dipper box, zone switching just after.

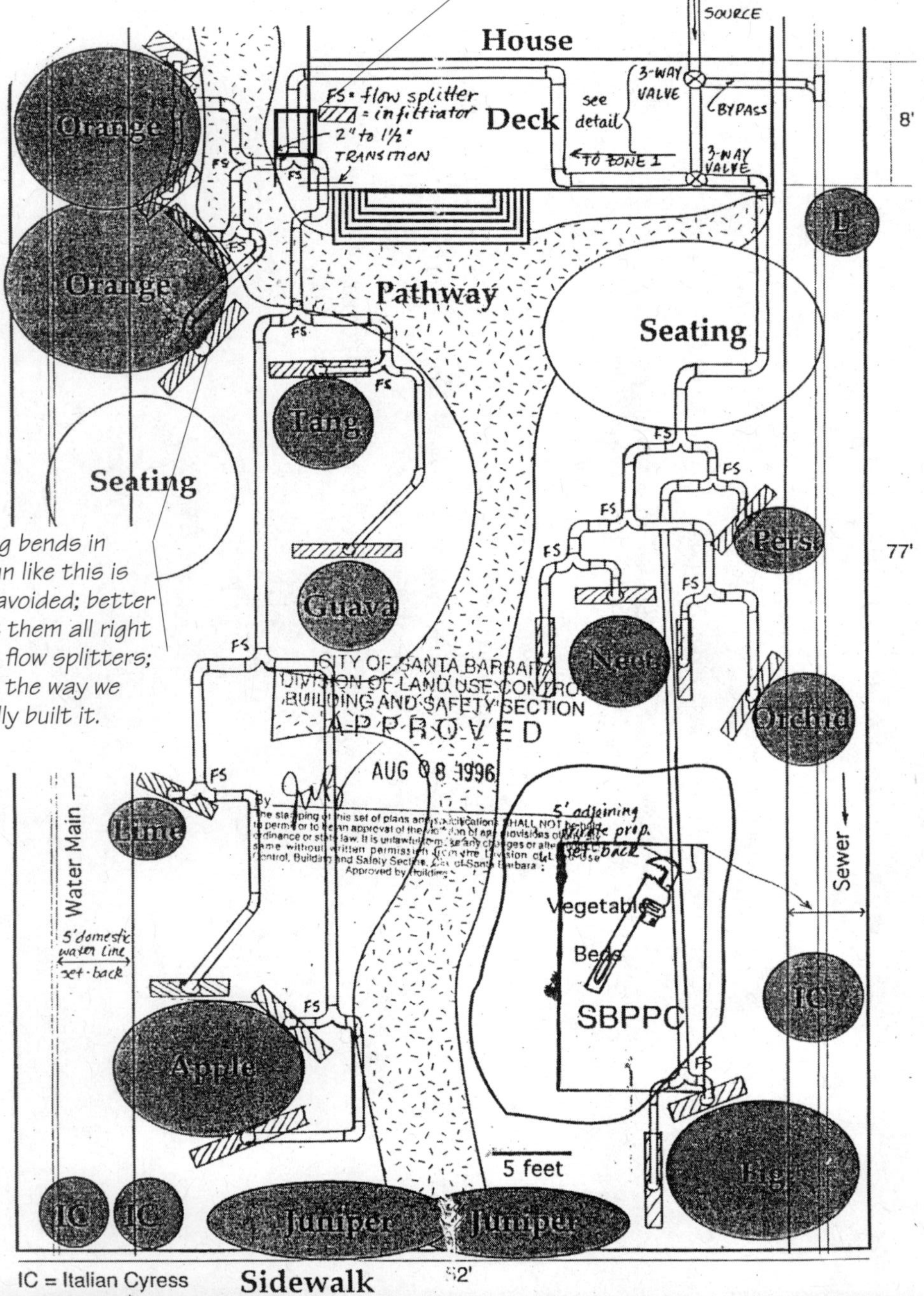

Putting bends in mid-run like this is to be avoided; better to put them all right by the flow splitters; that's the way we actually built it.

House

Deck

FS = flow splitter
= infiltrator
2"to 1½" TRANSITION

SOURCE

BYPASS

3-WAY VALVE

3-WAY VALVE

see detail

TO ZONE 1

8'

77'

Seating

Sewer

Pathway

Seating

Seating

Orange

Orange

Tang

Guava

Lime

Apple

Juniper

Juniper

Juniper

Persi

Orchid

Nect

Fig

Fig

Vegetable Beds

SBPPC

5 feet

5' adjoining
5' domestic water line
cut-back

CITY OF SANTA BARBARA
DIVISION OF LAND USE CONTROL
BUILDING AND SAFETY SECTION
APPROVED
AUG 08 1996

Water Main

Sidewalk

IC = Italian Cyress

IC

IC

44

Example 2: Urban Farm Bathhouse (Branched Drain, Composting Toilet)

These permit excerpts are for a bathhouse at the Center for Urban Agriculture at Fairview Gardens, an organic farm in Southern California. The Branched Drain to mulch basins greywater system plan was subjected to extra scrutiny because it had neither a sewer hookup nor a septic for treatment backup (see G-1f, G-9e, CPC). A composting toilet was part of the same permit, which took two years to get. A backup sewer connection would have cost several thousand dollars. The Center's good presentation and persistence in pioneering this approval cost them a great deal in delay and education for the Administrative Authority. However, they achieved important precedents in wise use of the Authority's discretion on the composting toilet and waiver of the redundant greywater backup.

Greywater System Approval Conditions

Submitted plans show infiltration area to be constructed initially, with dashed lines showing provision for doubling the infiltration area (see below).

The system to be installed with extra outlets on the distribution box which can supply additional infiltration area if this is ever required.

The efficacy of the flow splitters and the installed system capacity to be verified during inspection (alternatively, each infiltration area may be supplied with its own direct line from the distribution box).

a) The flow splitters will be visually inspected to verify that they are functioning correctly. If they are not, the capacity of the infiltration basins must be increased to be able to receive the total, unsplit flow from each line.

b) The system to be loaded with 240 gal of water in 1 hr (twice the expected daily loading) without water surfacing, overflowing, or the surface of the mulch becoming wet to pass inspection. If these conditions are not met, the installed infiltration area is to be doubled as per the submitted plans, using the extra outlets from the distribution box, and the test to be repeated. If the enlarged system does not pass the test, it will not pass inspection.

The inspector had not seen double-ell flow splitters in action and didn't know if they'd work.

TABLE 4: GREYWATER SYSTEM CALCULATIONS

Assumptions

"Sandy loam" type soil per perk test
Maximum absorption capacity = 2.5 gal/ft^2/day as per greywater code Table G2
Usage calculations used in place of bedroom-based calculations due to absence of bedroom
Measured shower flow rate was 1 gpm
4.6 gpm max flow rate with all fixtures on

The expected load is less than ¼ of the design load for the calculations. This is often the case with aggressive conservation. Especially when the legal load is "per bedroom," they are assuming from the outset you'll have 2-4x more water going into the system than you probably will. Always include projected actual flows on the calcs page to show the Administrative Authority that there is another huge margin of safety.

Greywater sources:

Showers/day/person	People	Min Rate (gpm)		Projected gpd
MAX load (conventional showerhead, maximum number of users)				
1	10	8	1.3	104
Projected actual load (ultra low flow showerhead in insulated bathing enclosure)				
1	10	8	0.3	24

Lav uses/day/person

3	10	0.5	1	15
Total greywater generation				
Max				119
Projected actual (not used, for information only)				39

Safety factors are not required information, but they put things into perspective. The system capacity is nearly 10x what is needed for the design load. The system can handle 1.63x the flow of turning all the fixtures on and leaving them on indefinitely. The system could handle twice the flow of all the fixtures running simultaneously for 1 hr. Any questions? (Note: The safety factors assume LTAR = initial perk rate; see p. 9.)

For the proposed "240 gal in 1 hr" test, about half would be filling the surge capacity and half absorbed. The theoretical 1 hr capacity is 558.6 gal.

Infiltration adds a few percent more.

Almost an entire day's flow can be accommodated without any infiltration.

This was the richest soil we've ever seen: a great asset for effective treatment.

Greywater distribution-48ft^2 infiltration area

portion of flow	gpd	area	gal/ft^2/day	% of 2.5 gal/ ft^2 permitted	# of outlets	gal. surge capacity (zero infiltration)*	% of daily flow	Infiltration Capacity gpd	1 hr surge to daylight effluent Perk (min/inch):
									5
Per 1/8th flow outlet									
0.125	14.9	6.0	2.5	99%	2	11.2	9%	1083	55.9
Per1/4 flow outlet									
0.2500	29.8	12.0	2.5	99%	4	22.4	19%	2165	111.7
total									
1.0000	119.0	48.0	2.5	99%	6	112.2	94%	10826	558.6
								8	9.31 gpm

9097%	163%		202%
Saftey factor over expected use	Continous max flow safety factor		1 hr max flow safety factor

I included the calcs for the system sized the way I wanted, and twice as big, per the approval conditions. In use, the tiny actual flow can be switched between two 24ft^2 ones which will have separate valves controlling backup drip irrigation.

Greywater distribution-96ft^2 infiltration area (extra capacity option)

portion of flow	gpd	area	gal/ft^2/day	% of 2.5 gal/ ft^2 permitted	# of outlets	surge capacity (zero infiltration)*	% of daily flow	Infiltration Capacity gpd	1 hr surge to daylight effluent
Four 1/16th flow outlets									
0.0625	7.4	6.0	1.2	50%	4	11.2	9%	1083	55.9
Six 1/8 flow outlets									
0.1250	14.9	12.0	1.2	50%	6	22.4	19%	2165	111.7
total									
1.0000	119.0	96.0	1.2	50%	10	179.5	151%	17321	893.8
								12	14.896 gpm

*adjusted for 50% mulch volume

14556%	261%		324%
Saftey factor over expected use	Continous max flow safety factor		1 hr max flow safety factor

FIGURE 8: URBAN FARM BATHHOUSE BRANCHED DRAIN TO MULCH BASINS DRAWINGS

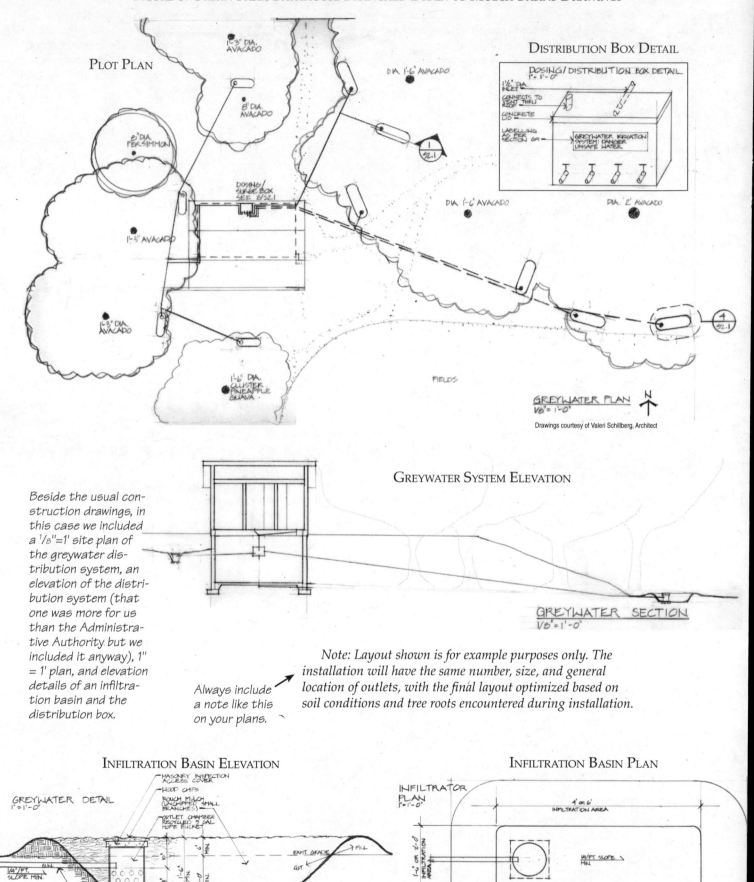

Beside the usual construction drawings, in this case we included a ¹/₈"=1' site plan of the greywater distribution system, an elevation of the distribution system (that one was more for us than the Administrative Authority but we included it anyway), 1" = 1' plan, and elevation details of an infiltration basin and the distribution box.

Always include a note like this on your plans.

Note: Layout shown is for example purposes only. The installation will have the same number, size, and general location of outlets, with the final layout optimized based on soil conditions and tree roots encountered during installation.

Drawings courtesy of Valeri Schillberg, Architect

46

Example 3: Have Vendor Deal with Permit
(Automated Sand Filter to Subsurface Emitters)

*If you buy a turnkey system, the vendor may deal with the permit for you. ReWater Systems,[e] for example, has gotten permits for several hundred systems under the CPC. Here's the **first page** of their five-page form to give you an idea of how they'd do it for you:*

ReWater® System Building Permit Application and Compliance Form
for Installation of a Graywater Irrigation System Under the California Plumbing Code

SUBMITTED FOR: John Waters   SUBMITTED BY: ReWater Systems, Inc.
                 123 Pines Terrace   477 Marina Parkway
                 Del Mar, CA   Chula Vista, CA 91910
                                                 (619) 585-1196
                                                 Contractor's License #798547

INTRODUCTION

ReWater Systems, Inc. sponsored the 1992 California greywater legislation, and we cooperated closely with the California Department of Health Services and Department of Water Resources during the three years it took them to write the greywater code, entitled Appendix G of the California Plumbing Code, heretofore the "Code".

We provide this form to help building inspectors understand the Code's requirements for issuing a building permit for a ReWater® subsurface drip greywater irrigation system.

GENERAL DESCRIPTION

THE REWATER SYSTEM - Mechanical

A ReWater system meets or exceeds all Code requirements. It collects greywater in a surge tank, filters it of suspended solids in a filter unit, then automatically distributes it via buried low pressure polyethylene tubing to multiple valved subsurface drip irrigation zones located throughout the landscape. Each valve is operated by a programmable controller. Each zone consists of numerous patented subsurface drip irrigation emitters manufactured especially for greywater and listed by the Center for Irrigation Technology as required by Section G-11(a)(2) of the Code.

THE PROPOSED SITE - Geological

In the process of completing this application, we determined that this site meets or exceeds all Code requirements. Code requires a minimum size greywater site, which is determined by the amount of water produced by the house, the type of soil at the site, and the setbacks found in Table G-1.

FORMAT

Code requirements are summarized in normal type face. Where the Code is quoted, it is surrounded by quotation marks ("). Where the Code requires some form of action by either the installer or inspector, words are typed in *italics*. Where the Code requires the manufacturer to supply information, that information is supplied and underlined.

Section G-1   Graywater Systems (General).

(a) Greywater is for subsurface irrigation only.
    The house water must be protected by a Reverse Pressure Principle Device.
    No surfacing of graywater is the intent of this code.
    Your local plumbing code remains the same except as hereby altered.
(b) All connected graywater is discharged into subsurface drip irrigation fields.
(c) The entire greywater system is confined to the lot of the discharging building.
(d) *A plot plan is hereby submitted* (Section G-4(a)), prepared by Z: Architecture
    No inappropriate soil conditions.
(e) Not located in a geologically sensitive area.
(f) Appropriate clearances have been met per Table G-1.
    This house is not on a private sewage disposal system.
(g) Operation and maintenance manual has been provided to the system owner by ReWater Systems, Inc.
    A copy is hereby attached.
(h) Administrative Authority must supply copy of Appendix G to permit applicant.

Section G-2   Definitions.

"Graywater is untreated household waste water which has not come into contact with toilet waste. Graywater includes waste water from bathtubs, showers, bathroom wash basins, clothes washing machines and laundry tubs, or an equivalent discharge as approved by the Administrative Authority..."

References and Suppliers

Caution: This list does not constitute an endorsement of suppliers or their offerings.

[1] **The New Create an Oasis with Greywater** Art Ludwig, 2006. Oasis Design, 5 San Marcos Trout Club, Santa Barbara, CA 93105-9726. oasisdesign.net Phone: 805-967-9956, Fax: 805-967-3229.

[2] **US Green Building Council** www.usgbc.org *Green Building certification, green building resources.*

[3] **Greywater Policy Center** oasisdesign.net/greywater/law

[4] **Common Greywater Mistakes and Preferred Practices** oasisdesign.net/greywater/mistakes

[5] **ReWater Systems Inc** P.O. Box 210171, Chula Vista, CA 91921. www.rewater.com Phone: 619-421-9121, Fax: 619-585-1919, E-mail: support@rewater.com. Contact: Steve Bilson. *Maker of plastic distribution cones and a range of GW systems from $1,295 and up. Active in GW politics.*

[6] **Cost/Benefit Spreadsheet** oasisdesign.net/downloads/rewatergwcalculator.xls *Example of a comprehensive cost benefit analysis from ReWater Systems.*[5]

[7] **Jade Mountain Inc./Real Goods** 360 Interlocken Blvd., Ste. 300., Broomfield, CO 80021. www.realgoods.com/renew Phone: 800-442-1972 (orders), Fax: 303 222-3599, E-mail: info@jademountain.com *Mail-order source for spa-flex, 3-way diverter valves, GW systems and parts.*

[8] **Infiltrator Systems Inc.** 6 Business Park Rd., P.O. Box 768, Old Saybrook, CT 06475. www.infiltratorsystems.com Phone: 800-718-2754, 860-577-7000, Fax: 860-577-7001, E-mail: info@infiltratorsystems.com. *Plastic infiltration chambers of good quality.*

[9] **NutriCycle Systems (Formerly Hanson Associates)** 3205 Poffenberger Rd., Jefferson, MD 21755. www.nutricyclesystems.com Phone: 301-371-9172, Fax: 301-371-9644. Contact: John Hanson, E-mail: jhanson@nutricyclesystems.com. *"Nutrient recycling system" composting toilet, box trough GW systems. Leaching chamber and box trough designs are thanks to John Hanson.*

[10] **Clivus Multrum Inc.** 15 Union St., Lawrence, MA 01840. www.clivusmultrum.com Phone: 800-4-CLIVUS (800-425-4887), 978-725-5591, Fax: 978-557-9658, E-mail: forinfo@clivusmultrum.com. *Longtime manufacturer and distributor of composting toilets. Supplier of GW systems and information.*

[11] **Geoflow Inc.** 506 Tamal Plaza, Corte Madera, CA 94925. www.geoflow.com Phone: 800-828-3388, 415-927-6000, Fax: 415-927-0120. Contact: Karen Fergson. *Their premium underground drip irrigation tubing is impregnated with herbicide to keep roots out and has a good reliability record.*

[12] **Orenco Systems Inc.** 814 Airway Avenue, Sutherlin, OR 97479-9012. www.orenco.com Phone: 800-348-9843, 541-459-4449, Fax: 541 459-2884. *Established makers of sand filter septic systems, components, seminars.*

[13] **Design Manual: Constructed Wetlands and Aquatic Plant Systems for Municipal Wastewater Treatment** US Environmental Protection Agency, Office of Research and Development, 1988. Center for Environmental Research Information, 26 W. Martin Luther King Dr., Cincinnati, OH 45268. Phone: 513-569-7562, Fax: 513-569-7566. *To order a copy, call 800-490-9198. You can get a pdf of the book at www.epa.gov/owow/wetlands/pdf/design.pdf.*

[14] **Tad Montgomery & Associates** P.O. Box C-3, Montague, MA 01351. Phone: 413-367-0068, E mail: tad@shaysnet.com. *Ecological Engineering, constructed wetlands, composting toilets, etc.*

[15] **Natural Systems International (Formerly Southwest Wetlands Group)** 3600 Cerrillos Rd., Ste. 1102, Santa Fe, NM 87507. www.natsys-inc.com Phone: 505-988-7453, Fax: 505-988-3720, E-mail: nsi@natsys-inc.com. Contact: Michael Ogden. *One of the first engineering firms specializing in constructed wetlands—a good outfit.*

[16] **Graywater Guide: Using Graywater in your Home Landscape** 1995. California Department of Water Resources, P.O. Box 942836, Sacramento, CA 94236-0001. www.owue.water.ca.gov Phone: 916-653-5791. Greywater contact: Julie Saare-Edmonds, Phone: 916-651-9676, E-mail: julieann@water.ca.gov. *Guide and revised greywater standards (1997) both available at the department's website.*

[17] **Center for Irrigation Technology** California State University-Fresno, 5730 N. Chestnut Ave., Fresno, CA 93740-0018. www.cati.csufresno.edu/cit Phone: 559-278-2066, Fax: 559-278-6033. *At the website, request "Irrigation Equipment Performance Report" for emitter ratings.*

[18] **Netafim USA** 5470 E. Home Ave., Fresno, CA 93727. www.netafimusa.com Phone: 888-NETAFIM (888-638-2346), 559-453-6800, Fax: 800-695-4753, 559-453-6803. *Manufacturers of subsurface drip tubing with mechanical barrier against root intrusion. Though not as root-resistant as GeoFlow, it is herbicide-free.*

[19] **Personal communications** Gerry Winant, Supervisor for Environmental Health, Santa Barbara County, CA; Bob Sedivy, Plan Check Supervisor, Santa Barbara City, CA.

[20] **Our own door to door survey** in Santa Barbara and Goleta, CA in 1990, at the height of a seven year drought.

[21] **Greywater Pilot Project Final Report** Los Angeles Department of Water Reclamation, 1992. May be available from Los Angeles Department of Water and Power, P.O. Box 51111, Rm. 1315, Los Angeles, CA 90051-0100. www.ladwp.com Phone: 213-367-4141, Fax: 213-367-0907. *Report on first quantitative field testing of greywater health safety.*

[22] **Excreta Disposal for Rural Areas and Small Communities** E. Wagner and J. Laniox, 1958 (reprinted 1971). Monograph #39, World Health Organization, Geneva, Switzerland. *Information and diagrams about the effectiveness of soils for containing and processing nutrient and pathogen contamination from human feces, as well as other technologies.*

[23] **Green Land, Clean Streams: The Beneficial Use of Waste Water through Land Treatment** Daniel J. Elazar, 1972. From page 176. Center for the Study of Federalism, Temple University, 1616 Walnut St., Rm. 507, Philadelphia, PA. 19103. www.temple.edu/federalism Phone: 215-204-1482, 215-204-7784, E-mail: v2026r@vm.temple.edu. *Out of print but photocopies available.*

[24] **Oasis Biocompatible Products, available from Bio Pac, Inc.** 584 Pinto Ct., Incline Village, NV 89451. www.bio-pac.com Phone: 800-225-2855, Fax: 866-628-1662.

Further Reading

Using Greywater Carl Lindstrom, Clivus Multrum Inc.[10]

The Chemical, Physical and Microbiological Characteristics of Typical Bath and Laundry Waste Waters W.D. Hypes, C.E. Batten, and J.R. Wilkins, 1974. Langley Research Center, NASA. www.ntrs.nasa.gov *Accessible at NASA Technical Reports Server (NTRS) website via search.*

"Assessment of On-Site Graywater and Combined Wastewater Treatment and Recycling Systems Report" National Association of Plumbing-Heating-Cooling Contractors, 180 S. Washington St., P.O. Box 6808 Falls Church, VA 22046. www.phccweb.org Phone: 800-533-7694 703-237-8100, Fax: 703-237-7442, E-mail: naphcc@naphcc.org.

Greywater chemistry information www.greywater.com

A Cycle of Cycles Guide to Wastewater Recycling in Tropical Regions Katja Hansen and Douglas Mulhall, 1998. European Commission, Directorate General 1-B, Hamburg, Germany. *Economic, social and technical guide to tropical and Mediterranean wastewater recycling for flows from less than 50,000 inhabitants, with special reference to agriculture.*

Domestic Greywater Reuse: Overseas Practice and its applicability to Australia Barry Jeppesen and David Solley, Brisbane City Council. Research Report No. 73, March 1994. Urban Water Research Association of Australia, c/o Melbourne Water, P.O. Box 4342, Melbourne VIC 3001, Australia. www.melbournewater.com Phone: +61-3-9235-7100, Fax: +61-3-9235-7200. *A wealth of information of greatest interest to regulators. There is relatively little here for the individual or contractor. A download of the paper can be purchased at www.wsaa.au, via the bookshop link.*

Index

Also from Oasis Design...

❖ *The New Create an Oasis with Greywater* Why to use/not use greywater, health guidelines, greywater sources, irrigation requirements, 20 system examples and selection chart, biocompatible cleaners, greywater plumbing principles, components, maintenance, troubleshooting, freezing, rain, preserving soil quality, storing rainwater, suppliers, further reading. 146 pages, 53 figures, 130 photos, $20.95.

❖ *Greywater Extras* Electronic files with greywater calculation spreadsheets, subtropical and low chill deciduous fruit tree chart, New Mexico plant list. See oasisdesign.net/greywater for specs and price.

❖ *Principles of Ecological Design* Natural living is the harmonious integration of human culture, technology, and economics with nature. This booklet explains principles for redesigning our way of life to live better with less resource use. 18 pages, 30 figures and photos, color download $4.95, or B&W hard copy, $6.95.

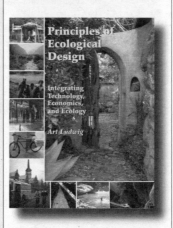

❖ *Water Storage* Do-it-yourself guide to designing, building, and maintaining water tanks, cisterns, and ponds, and managing groundwater storage. It will help you with your independent water system, fire protection, and disaster preparedness, at low cost and using principles of ecological design. Includes how to make ferrocement water tanks. 136 pages, 43 figures, 128 photos, $19.95.

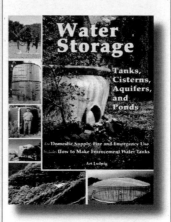

❖ *Laundry to Landscape Greywater Systems DVD* An instructional video on how to design and build Laundry to Landscape Greywater Systems. These are the simplest, most economical greywater systems to install yourself as an owner or renter, and a good green collar job opportunity (as a self-employed installer or as part of your landscape, plumbing, or construction business). 90 minute DVD video, soy ink printed in a recycled cardboard case, $19.95

...and more...
See oasisdesign.net/catalog

Oasis Design Consulting Service

A consultation is a great way to optimize your system design and ensure it fits well with your water supply, landscape, and energy systems. We can design these other systems also. Most design consultation can be done inexpensively off-site. Check our website (oasisdesign. net/design/consult) or contact us for more information.